WAITING

FOR

THE

SKY

TO FALL

Ruth Mackay

WAITING FOR THE SKY TO FALL

THE AGE OF VERTICALITY IN AMERICAN NARRATIVE

The Ohio State University Press | Columbus

Copyright © 2016 by The Ohio State University.
All rights reserved.

Library of Congress Cataloging-in-Publication Data
Names: Mackay, Ruth, 1985– author.
Title: Waiting for the sky to fall : the age of verticality in American narrative / Ruth Mackay.
Description: Columbus : The Ohio State University Press, [2016] | Includes bibliographical references and index.
Identifiers: LCCN 2016012405 | ISBN 9780814213087 (cloth ; alk. paper) | ISBN 0814213081 (cloth ; alk. paper)
Subjects: LCSH: American fiction—20th century—History and criticism. | American fiction—21st century—History and criticism. | Place (Philosophy) | Space and time. | Metaphor.
Classification: LCC PS369 .M33 2016 | DDC 810.9/005—dc23
LC record available at https://lccn.loc.gov/2016012405

Cover design by Susan Zucker
Text design by Juliet Williams
Type set in Adobe Sabon

∞ The paper used in this publication meets the minimum requirements of the American National Standard for Information Sciences—Permanence of Paper for Printed Library Materials. ANSI Z39.48–1992.

9 8 7 6 5 4 3 2 1

CONTENTS

List of Illustrations vii
Acknowledgments ix

INTRODUCTION The Age of Verticality: The Closure of the Frontier to the 9/11 Memorial, 1890–2011 1

CHAPTER 1 "Down Down Down He Shot": Winsor McCay's Work, Vertical Collapse, and Time in the Modernist City 32

CHAPTER 2 Upton Sinclair's Vertical Infernos: Oil Procurement and Disaster Culture 79

CHAPTER 3 "The Horizon Was an Illusion": Flight, Escape, and Imagining Vertical Space in Leslie Marmon Silko's *Almanac of the Dead* 111

CHAPTER 4 Terror/Power: Allen Ginsberg's Nuclear Poetics and the Space Race 147

CHAPTER 5 Traversing Vertical Space: Philippe Petit's Wire-Walk, Danger, and Transformation 180

CODA Up and Down Stories 210

Bibliography 221
Index 247

ILLUSTRATIONS

FIGURE 1   The South Pool of the 9/11 Memorial, New York   17
FIGURE 2   *Dream of the Rarebit Fiend,* Winsor McCay, 1906   37
FIGURE 3   The first of Winsor McCay's *Little Nemo in Slumberland* series, published in the *New York Herald,* 15 October 1905   39
FIGURE 4   *Little Nemo in Slumberland,* Winsor McCay, 29 September 1907   59
FIGURE 5   *Little Nemo in Slumberland,* Winsor McCay, 18 April 1909   65
FIGURE 6   *Little Nemo in Slumberland,* Winsor McCay, 27 January 1907   67
FIGURE 7   *Little Nemo in Slumberland,* Winsor McCay, 23 September 1906   73

ACKNOWLEDGMENTS

This book began in 2001. Like many others working in literary and cultural studies, I was influenced by the manner in which 11 September began to shape the discourses and narratives that came after it. I was a teenager when 9/11 occurred, and the event would in many ways determine my formal study of American narrative from the beginning of my university life. The work included in this book has been with me for a long time, and it has been shaped and refined with the help of many individuals who may not realize the ways that they have supported me.

I owe a great debt of gratitude to Jay Prosser and Andrew Warnes, both of whom provided me with great intellectual generosity and indefatigable good humor during the early stages of this book. They gave me gentle guidance for which I am very grateful.

I am very thankful to Stuart Murray for his great support and friendship, particularly during the early part of my work on this book. I am grateful to Hamilton Carroll and Catherine Morley, whose questions about my work helped me enormously in revisions. I appreciate their enthusiasm and encouragement as well as their specific advice in the early stages of putting this book together. I am also grateful to Nick Selby for his support of my early work on *Man on Wire,* and Bridget Bennett for the ways in which she helped me develop my methodology.

Stephen Mitchell has provided me with many long, inquisitive, and enriching conversations on cinema and representation. Writing about *Man on Wire* with him helped me in my own revisions of chapter 5 of this book. I am grateful to him for his enthusiasm and his reminder that it's normal to admit that you don't fully understand Derrida, as well as his willingness to discuss that lack of understanding.

Thank you to Lindsay Martin at The Ohio State University Press for her astute, wise, and practical advice throughout the process of redrafting my book. I appreciate her support, guidance, and willingness to help me produce the best possible work.

The peer reviewers who kindly gave full and detailed reports of my manuscript were insightful, and I greatly appreciate their thoughts about how to improve my work.

I am grateful to Taylor & Francis for permission to reprint parts of a previous article, "'Going Backwards in Time to Talk about the Present': *Man on Wire* and Verticality after 9/11," which was published in *Comparative American Studies* 9:1 (March 2011): 3–20 (available at http://www.tandfonline.com).

The Arts and Humanities Research Council and the University of Leeds in the U.K. provided me with opportunities to continue my work, for which I am very grateful. The University of Leeds offered me several scholarships that enabled me to pursue my postgraduate work, and the AHRC provided funding for the completion of that work.

The staff at the British Library in Boston Spa have been kind and helpful to me during every visit I've made to the branch. Sara Duke at the Library of Congress provided me with generous assistance and expertise during a research trip to view Winsor McCay's work in 2011. I also appreciate Johnny Gore's help in locating and viewing the films of Winsor McCay held by the New York Public Library during the same trip.

The support of friends and family through the writing of this book has been instrumental. I appreciate, as ever, the love and support of my brother Ian, Sarah and Mark Bradley, Romy and Callum Bundy, Julia Bullock, Judith Ernst, Meghan Wright, Annalise Ford, Claudia Troiano, and Rima Salloum. The conversations about literature I had with Kelly Quindlen and J. J. Pettijohn in the period of redrafting my work helped me enormously, though they may not realize it, and I continue to appreciate their friendship.

I owe my parents, Margaret and Christopher Mackay, so much. Without their continued support and love, I would have been unable to produce what follows, and they have provided me with both the practical tools and the emotional support to write this book. I am ever grateful for their belief in me, and for this and everything else I thank them deeply.

I am always indebted to Theresa D. Smith for reminding me how beautiful literature can be. Her insightful notes and sharp editing have made this book much more fluid than it would have been without her, and she has enriched my work in infinite ways. She has given me so much, but her gentle and unwavering support through the process of writing this book is an expression of love that has allowed it to exist.

This book is dedicated to my family.

INTRODUCTION

THE AGE OF VERTICALITY

The Closure of the Frontier to the 9/11 Memorial, 1890–2011

This book is about the cluster of vertical events, facts, and representations that preoccupy American narrative from the end of the twentieth century into the post-9/11 moment. It is about witnessing vertical moments, or imagining entirely new ones, and the representations that emerge in prose, poetry, comics, photography, and film. The texts discussed in the following chapters are not always about looking up, but they repeatedly attach verticality to the sense of trepidation denoted by the idea of waiting for the sky to fall. This book's organizing phrase, then, refers to verticality imbued by imminence and anxiety: a set of representations that worry at an unknown future or a not-well-understood present. The chapters that follow explore a literary, visual, and physical genealogy of verticality, beginning with readings in this introduction of post-9/11 prose and physical objects like the 9/11 Memorial and One World Trade Center, and ending with a genre of post-9/11 texts that includes memoir, documentary film, and visual narrative. Together, the chapters that follow ask how verticality after 9/11 crystallizes the affective dispositions of narratives across the long twentieth century.

Waiting for the Sky to Fall begins in 1890, the year that the U.S. Census declared the end of unsettled land lying to the west. The end of the frontier signaled radical changes in ways of imagining space, marking the end of what had previously been imagined to be limitless horizontal land west of the nation. The specifically American use of the term *frontier* does not identify a static and permanent border between two nations but rather "an area inviting entrance"; the "transient and temporal" space to the west of Jamestown, Virginia, into which Euro-American civilization expanded after 1607.[1] The space imagined to be conquerable moved horizontally across the nation and was also conceived as immense; Thomas Jefferson and his peers proposed that it would take a thousand years to reach the Pacific.[2] Westward expansion was in this sense not only a physical movement but also a crystallization of the belief that the new nation held a nearly limitless natural bounty. When the U.S. Census declared the end of the frontier, it became clear that the landscape had not been as vast as Jefferson had imagined. While 1890 hardly signaled the end of American aspirations for conquering space, energies of expansion could no longer extend horizontally. In this book, I use the term *vertical frontier* to signify the vertical spatial metaphors that replaced horizontal ones after the closure of the frontier. I argue that, facing the loss of what was imagined to be endless westward land, narrative attention turned toward the verticality of skyscrapers, aviation, drilling, and mining. If verticality did not entirely replace the horizontality of westward expansion, it did provide a new, and less constrained, metaphor for cultural texts.

The two events that comprise the foundation to this book reveal how larger conversations are transported by spatial metaphors. Both the closure of the frontier in 1890 and the fall of the Twin Towers in 2001 are events that leak into narrative in interestingly oblique ways, offering a transformed concept of the limits of space. The events of 11 September 2001 were lucidly vertical. They were also devastatingly felt by many individuals, making it important to acknowledge the many griefs that occurred on that day and in its aftermath. But it is also impossible to talk about the 2001 attacks without direct reference to the planes that descended from the sky, the enormous towers that were ruined in doubled moments of vertical collapse, and the bodies of those individuals who either fell from the Twin Towers or, facing the horror inside the buildings, chose to jump from them. As this book explores, images of falling and vertical ruin are

---

1. Webb, "American Frontier Concept," 3–5.
2. See Jefferson, *Notes on the State of Virginia*. See also Trachtenberg, *The Incorporation of America*, 11ff.

not confined to the post-9/11 text, but the unavoidable verticality of 9/11 ensures that narratives concerned with representing either that day or its aftermath must grapple with residual images of bodies falling and towers disintegrating. The frontier's closure removed a sense of physical horizontal space, producing an eruption of vertical metaphor; the event marked the beginning of what emerges in the following chapters as a complex triangulation between specific historical events, imagined moments, and narrative representation. As I discuss in detail in later parts of this introduction, the events of 11 September 2001 fully reveal this process: in the post-9/11 moment, vertical representation cleaves to, and often transforms the contexts and associations of, specific events that are physically and visually disorienting, disquieting, or even traumatic.

While the frontier was officially "closed" in 1890, it was at the Chicago World's Fair in 1893 that Frederick Jackson Turner delivered his speech, "The Significance of the Frontier in American History." Turner argued that westward expansion had determined the cultural, legal, and political characteristics of the nation, and his speech marked an explicit, public, declarative end of horizontal space and an argument for the full significance of that same spatial construct. The Chicago World's Fair also fully exposed an all-consuming desire among American engineers and architects to build ever higher into the sky, a continuation of the tall-building trend that started its life in the same city in the 1880s. In 1889, the Eiffel Tower—unveiled at the Paris World's Fair—had surpassed the Washington Monument to become the tallest man-made structure in the world. George Washington Gale Ferris's Wheel was one of the greatest achievements of the Chicago World's Fair: a vertical feat that was brought into existence by the calls from the event's planners for something that would "out-Eiffel Eiffel."[3] The year 1890, then, is the beginning of this book for two reasons: the "end" of horizontal space in the United States and the beginning of an American imperative for verticality.

This book does not prioritize vertical imagery that refers to a particularly established or recognizable idea such as flight or falling, although it frequently distinguishes between *concrete* or *abstract* verticality. The former refers to particular incidences in which verticality is an integral part of the event's cultural or technological significance. Concrete verticality includes moments defined by a specific vertical spectacle, such as the first successful flight of a human-propelled airplane by the Wright brothers in 1903, the dropping of atomic bombs from planes over Japan in 1945, and the collapse of the Twin Towers in 2001. It also includes events that

---

3. Larson, *Devil in the White City*, 134.

leave behind a tangibly vertical structure such as the building of the first American skyscrapers in Chicago in the late 1870s, the vertical tunneling of mine shafts in the last years of the nineteenth century, and the erection of oil derricks in the first years of the twentieth century. By contrast, abstract verticality refers to those images that do not conjure up a tangible event nor allude to a particular historical moment, but rather hinge upon intangible ideas. A primary example of abstract verticality is the concept of "being high" as a spatial metaphor that conflates the idea of being in an altered state of mind with a physical location.

Verticality is a constellation of imagined ups and downs, discrete metaphors that exist along a linear axis, and larger stories of gravity. Verticality more specifically denotes preoccupations with flight, real or imagined, literal or figurative; the problematic, traumatic, and spectacular fall; the erection of buildings and structures extending into the sky; the dynamics of tunneling into the earth or standing on a precipice and looking down at a world below. This book traces an expansive genre not limited to only one direction—up or down—because the ways in which verticality draws its force are, importantly, multidirectional. In the following chapters, I work toward a comprehensive understanding of verticality by employing a methodology that pays attention to the specific characteristics of key narratives across the twentieth century. The textual representations discussed in this book are united by a return to the effects of a sensory experience concerning an overwhelming vertical event or structure. However, these same representations do not only mediate what is visually vertical: they frequently either reimagine those same moments, or imagine entirely new ones. This results in moments that may be directly prompted by a precise vertical event but that inevitably become part of the wider, and more complex, process of unraveling the route between the physical and nonphysical.

The historical context for verticality is as rich as its referents. Vertical metaphor draws its power, in part, from the connotations of the vertical axis: a cultural taxonomy that polarizes up and down as well as the subjective ideas attached to these positions. This taxonomy finds its precedents in the Judeo-Christian model of heaven and hell and the Renaissance concept of a Great Chain of Being. The vertical axis cleaves positive ideas such as morality, illumination, and wealth to height; conversely, what is underneath or underground becomes a source of degradation, darkness, and poverty.[4] Verticality is a category of representation that has

---

4. My model of the vertical axis here is informed by Lovejoy, *Great Chain of Being*; van Leeuwen, *Skyward Trend of Thought*, 99–101; and Kern, *Culture of Time and Space*, 242–47; 310–12; 317. Studies in similar areas include Falconer, *Hell in Contemporary Literature*; Bachelard, *Air and Dreams*, 10ff.

significance to intersecting fields of study in different genres, but previous studies have focused only on a specific aspect of verticality, such as falling, flight, or underground spaces, and tend to refer to the vertical axis as the governing context for vertical representation. An exception comes with Kristen Whissel's study of digital effects in blockbuster cinema since the 1990s, which does gesture toward the significance of verticality after 9/11; Whissel notes that, after 2001 "the (hauntingly vertical) photograph of 'The Falling Man'" operates as a contemporary analogue to the figures of Phaeton and Icarus that occurred in emblem books—"ready to accommodate a range of contradictory and even oppositional meanings."[5] Whissel's concern, like mine, is with "spatial dialectics," but her interest is directed at contemporary cinema and its "vertical imagination,"[6] rather than the roots of such a tendency in American narrative and its emergence in post-9/11 narrative.

Verticality often imagines or directly represents disastrous moments and provides a recurring, though sometimes nebulous, sense of unknowing. The age of verticality begins at a point when revolutionary changes involving the construction of the skyscraper and the advent of aviation were occurring. But vertical imagery across the twentieth century transgresses the boundaries of discourses surrounding individual vertical events, instead becoming an intrinsic part of narratives that are not merely related to the material and visible verticality of the skyscraper or the airplane. This book's interdisciplinary bent is the result of verticality's overflow into categories of abstract and concrete; using the insights of an expansive field of theory provides the best way of understanding such a broad field as vertical imagery. In its course, this book claims verticality as a transformative category that uses metaphor to untangle, interrogate, or underscore specific epistemological and philosophical ideas.

TRANSFORMATIVE VERTICALITY: HENRY ADAMS'S DYNAMO AND PHILIPPE PETIT'S TWIN TOWERS

The period of burgeoning verticality at the fin de siècle is uniquely refracted by Henry Adams's *The Education of Henry Adams,* which yokes the vertical moment to concepts of visuality, time, and what is material or abstract. *Education* is a third-person narrative permeated by "Henry Adams" the protagonist and Henry Adams the author, a painstakingly

---

5. Whissel, *Spectacular Digital Effects,* 57.
6. Ibid., 22.

constructed text in which the autobiographical figure is "always subjected to historical forces both more powerful and complex than any he can muster as mere intellectual."[7] The narrative is a cyclical examination of Henry Adams's repetitive "failure" to obtain the education he feels is required for a proper understanding of the world, but its climax is a meditation on the years around the turn of the century that culminates in Adams's "dynamic theory of history."[8] The author's expression of awe in the presence of the forty-foot-tall dynamos is part of his study of "the forces" conducted in the closing years of the twentieth century (317).

Adams "haunted" the Paris Exposition of 1900 until its closure and, under the tutelage of Samuel Pierpont Langley (an astronomer and aerodynamicist), observed the new Daimler motor and automobile, and then the dynamos (317–18). The dynamos lead the narrative to the following exclamation:

> Satisfied that the sequence of men led to nothing and that the sequence of their society could lead no further, while the mere sequence of time was artificial, and the sequence of thought was chaos, [Adams] turned at last to the sequence of force; and thus it happened that, after ten years' pursuit, he found himself lying in the Gallery of Machines at the Great Exposition of 1900, with his historical neck broken by the sudden irruption of forces totally new. (320)

The dynamo, a machine that makes the generation of electric energy visible, provides the spectacle that precipitates Adams's most perturbed reflections on time and sequence. The revelation that the dynamo engenders—one in which both history and Adams's neck, conflated into one image, are broken—is ordered by Adams's physical perspective looking up at the structure. Moreover, the depiction of the dynamo entangles its physical verticality with the underlying effect scientific progress has had on abstract principles of time and energy.

It is not coincidental that the 1893 Chicago World's Fair, a scene of verticality and site of overwhelming demonstrations of newness, was the occasion that caused Adams to write: "Here was a breach of continuity—a rupture in historical sequence!" (286). The visual exhibitions of progress at the Chicago World's Fair in fact provoke repeated allusions to

---

7. Rowe, introduction to *New Essays on the Education of Henry Adams*, 6.
8. Adams, *Education*, 37; 395–406. All subsequent citations come in the main text in parentheses.

the idea of broken temporality beyond this suggestion of a "breach" and "rupture" in linear, continuous time. Later in his *Education,* Adams traces "the motion of thought" through history and asserts that the acceleration and direction of this thought have changed over time, but that these fluxes have not previously breached a general "continuity" (382). The turn of the century, however, marks the point at which this continuity "snapped" (382); at this moment "a new avalanche of unknown forces had fallen" that "required new mental powers to control" (384). Adams's overarching claim is that, between 1893 and 1900, humankind "translated" itself into "a new universe which had no common scale of measurement with the old" (319). The dynamos, the focal point for Adams in both Chicago and Paris, facilitate the author's understanding that the network of new ideas and technologies the dynamo stands for has opened a chasm between the past and the future.

Adams's *Education* specifically exposes how a vertical fact prompts a new vision, here one related to the scientific discoveries that promise, for him, the sense that "something new and curious was about to happen to the world" (284). Adams's declaration that something "new" is about to happen is then reiterated in the assertion that: "one lingered long among the dynamos, for they were new, and they gave to history a new phase" (287). This atmosphere of newness signals an anticipation, implicitly looking forward in time. But then, looking up at the dynamo, time almost seems to stop: Adams feels as though the traditional measure of time—the planet's "annual or daily revolution"—is now less significant; there is a new time forged by the dynamo's turnings:

> To Adams the dynamo became a symbol of infinity. As he grew accustomed to the great gallery of machines, he began to feel the forty-foot dynamos as a moral force, much as the early Christians felt the Cross. The planet itself seemed less impressive, in its old-fashioned, deliberate, annual or daily revolution, than this huge wheel, revolving within arm's-length at some vertiginous speed. (318)

In this almost paradoxical depiction of expectancy and time's normal passage being thwarted, the machine also comes to symbolize an interrupted expanse of time condensed by its symbolic referent: "infinity."

Adams's prose underscores the promise of an unknown future, even while the act of looking up at the vertical structure results in a textual flattening out of time itself in the assertion that the dynamo is a symbol of infinity. To Adams, the powerful man-made implications of the dynamo coalesce in the combination of its height and its revolving speed, the latter of which is crucially couched in relation to the sensation of vertigo. *Vertigo* (like *convert, verse, universe,* and *versatile*) shares the same root as *vertical: vertere,* the Latin for "to turn." *Vertical* itself derives from the early French *vertical* or Latin *verticalis,* both of which mean, as noted in *Merriam-Webster's,* "relating to a vertex, at the highest point overhead," from the earlier Latin *vertex,* meaning "top, highest point, pole (turning point), whirl."[9] This etymological legacy indicates something about the avenues that "verticality" as a concept takes: the shades of meaning in *vertical* cleave "summit" to a point of dramatic change (a turning point) or a scene of tumult (whirl). These are ideas that emerge throughout this book: verticality is repeatedly, even stubbornly, the scene of extremity and change, the site of confusion and often, in turn, revelation.

Adams's encounter with the spectacle of the dynamo, then, both appears to arrest time and also makes its vertical prominence into a symbol in which the slippage of time is essential. The prose's temporal oscillations, predicated on Adams's physical position below the dynamo, forge the act of looking up into a symbolic gesture toward the author's own anxieties about the future of America. Sanford Pinsker suggests that, for Adams, the dynamo is "a collective symbol for his sense that the past was irretrievable, the present chaotic and confusing, and the future a cause for deep concern" because, essentially, it "confounded every organizing principle that Adams had searched for."[10] Furthermore, the sequence depicting the dynamos undergirds *Education*'s final exposition of a "dynamic theory of history," in which historical sequence is imagined as the consequence of myriad interacting forces. Foreshadowing this model, *Education* suggests that the dynamo has destabilized the "historical mind" that previously could "think only in historical processes" and now was "helpless before a mechanical sequence" (287). Producing a kind of kaleidoscope through which Adams's previous conceptions of space and time now appear as a disorienting mosaic, the dynamo becomes an authoritarian power in the shadow of which even the planet itself loses its impressiveness. By making visible what was only theoretical at the advent of electricity, the dynamo produces a visceral experience of verticality while tying this experience to its consequences: namely, the various and unpredictable effects of what this new technology might engender.

---

9. See *Merriam-Webster's Intermediate Dictionary.*
10. Pinsker, "Henry Adams at Ground Zero," paragraph 7 of 21.

By fixing upon the dynamo's motion and its ominous silence—it is "barely murmuring"—Adams makes the physical immediacy of what he witnesses into the source of its transformative power; the author is bodily affected by the vision before him and also made to question its consequences for the unseen principles of science, the past, and the future. The hold of the dynamo over Adams lies in its disruption of the previous scientific tenets he has held to be true. In 1890, William James opined that while "each one of our various *ologies* seems to offer a definite head of classification for every possible phenomenon of the sort which it professes to cover," due to the obstinacy of the human mind, "when a consistent and organized scheme of this sort" has been "comprehended and assimilated," a different scheme is "unimaginable": no alternative "can any longer be conceived as possible."[11] The response of discomfort Adams has to the new scheme of technology contained by the dynamo, while embodying the same undesired mental obstinacy of James's description, does not cause Adams to excoriate the science involved. It nonetheless indicates the way in which vertical imagery operates as an attempt to make an abstract scientific concept in some way visible or tangible, and thereby work through the questions those abstract ideas might pose.

Adams's depiction prompts the question of whether it is primarily the spectacle of verticality that engenders wariness, in a manner akin to what is at stake in the representation of the vertical after 9/11. Like many post-9/11 narratives that would come a century later, Adams's prose is troubled by what Sacvan Bercovitch describes as the feeling that America is "rushing toward self-destruction in an entropic inversion of the work of redemption."[12] I am not the first to tie together Adams's meditation with either post-9/11 discourse or the other significant dialogues that produced a sense of seeing things anew in 1900 (most prominently, Max Plank's quantum theory and Sigmund Freud's *The Interpretation of Dreams*).[13] But this book interrogates how the individual fields in which these moments operate—physics, architecture, psychology, cultural history—are interdependent. Verticality repeatedly emerges in texts preoccupied by changes in representing and conceiving the world, whether in terms of new visual forms such as photography and cinematic film, the various theories of early psychology and psychoanalysis, or new scientific endeavors that enabled the splitting of the atom. The following chapters

---

11. James, "Hidden Self," 361.
12. Bercovitch, *American Jeremiad*, 195.
13. Pinsker, "Henry Adams at Ground Zero."

contend that verticality across the twentieth century frequently responds to various spectacles with a revelatory moment that is both engendered by contemporary, and also anticipates future, scientific progress.

The way that time is represented in relation to verticality in Adams's *Education* exemplifies a focus of this entire book's discussions. Throughout, I explore the ways in which a sense of anticipation or imminence determines verticality, producing new ways of understanding time. Adams's perspective from the base of an enormous dynamo cleaves the act of looking up to a transformed, and transformative, concept of time. In 1974, Philippe Petit stood at the top of the World Trade Center and looked into the gulf below, contemplated his imminent walk across a wire strung between the two towers, and concluded that: "Up here, time has lost all sense."[14] Petit's thoughts, taken from his memoir of the walk between the Twin Towers, complicate the forward thrust of time in more than one way. Their semantic weight is comparable to that of Adams's reflection upon an entirely new vision of the future. Both are prompted by a dizzying perspective—Adams's caused by the "vertiginous" revolutions of the dynamo above him; Petit's secured by the view from the then-tallest structure in New York. And, like Adams, Petit suggests time itself is disturbed by vertical extremity. But Petit's retrospective account after 9/11 also signifies a representative act that complicates time through its very existence; revivified by his own memoir and the film *Man on Wire* in the years after 2001, the representation of Petit's walk imaginatively rebuilds the no-longer-standing World Trade Center. The walk's representation is therefore not only about the transformation of space in its recollection of the Twin Towers. The walk at 1,350 feet high was inherently determined by the potential for disaster, and its representation after the fact of 9/11 reworks the straightforward linearity of time, diffusing the image of collapsing towers with a walk over them still intact. The way that representations of Petit's walk negotiate the vertical events of 2001 is key to this book's exploration of the profound significance of a relationship between verticality and 9/11.

The particular case of the falling body is one that cannot seem to be disassociated from its 9/11 frame of reference.[15] As Aimee Pozorski writes, no matter what "images or words or symbols that we use to communicate

---

14. Petit, *To Reach the Clouds*, 158.

15. For an insightful and thorough discussion of "the vexed image of the falling man" and the critical context of representing those who fell from the Twin Towers, see Pozorski, *Falling after 9/11*, especially 10–14.

in this post-9/11 world," they will "perpetually, inevitably, unwittingly point back to a contemporary American trauma we cannot seem to get out of our consciousness."[16] Ultimately, vertical imagery after 9/11 condenses the potentially fraught convergence between concrete and abstract verticality. Texts must contend with the litany of actual vertical traumas that occurred. They must also negotiate various abstracted ideas, such as those previously attached to the American passenger plane, nominated as "indigenous mobility and zest" and "the galaxy of glittering destinations" by Martin Amis in an article for *The Guardian* on 18 September 2001 titled "Fear and Loathing." Richard Gray reflects explicitly upon this process when he suggests that the falls of the Twin Towers and people from them have become "a powerful and variable visual equivalent for other kinds of fall."[17] Post-9/11 texts that subscribe to (or indeed reject) the belief that various structures such as those of American foreign policy disintegrated on 11 September 2001 along with the World Trade Center also participate in the creation of abstract verticality. The most famous example is perhaps Jean Baudrillard's assertion of "the symbolic collapse of a whole system," in which the image of an invisible epistemic disintegration is fastened to the image of the World Trade Center itself.[18] The parameters of this book are usefully defined by the way that narrative dialogues with the visible and physical manifestation of verticality (such as the events of 11 September 2001) as well as abstract ideas of ideological or political collapse.

In the following discussions, I explore how 11 September 2001 offers a prismatic key to verticality across the long twentieth century. The events of that day operate as historical fact and become the site of multiple acts of representation. And verticality is a modality that reveals the anxieties most clearly represented by post-9/11 texts. While each chapter of this book reads vertical metaphors and images in a text or set of texts that exemplify a moment in the twentieth century, the following chapters climax with a discussion of Petit's wire-walk, a return to the post-9/11 moment, and the consideration of how verticality in its long view can be brought to bear with great usefulness on 9/11 studies as a field. And taken together, the discussions of this book complicate the avenues established by 9/11 theory by asking why verticality is infused with the disruptive tendencies of traumatic time.

---

16. Pozorski, *Falling after 9/11*, 2.
17. Gray, *After the Fall*, 7.
18. Baudrillard, *Spirit of Terrorism*, 8.

IMAGINING UP AND DOWN AFTER 9/11: INVERTED SKYSCRAPERS AND THE 9/11 MEMORIAL

In the following, I consider two texts: Jonathan Safran Foer's 2005 novel *Extremely Loud and Incredibly Close* and the 9/11 Memorial Museum that was unveiled in 2011 at the site of the terror attacks in Lower Manhattan. Reading these two texts, I discuss in detail how vertical imagery negotiates the various and often conflicting memories of physically vertical events that have occurred, as well as the abstract ideas and qualities that are routinely attached to vertical spaces or movements. Together, the readings of Adams's work and post-9/11 texts that provide the basis for this introduction exemplify the entire book's trajectory. The end of the twentieth century begins a period in which being confronted by physical verticality overlaps with questions of temporality, and Adams's account ushers in scenes of textual disruption that are mediated by a sense of dissonant or fractured time in a way that prefigures the texts discussed at length in the chapters that follow. Just as verticality at the beginning of the twentieth century is infused by trepidation, often concerning what vertical structures might signify about technological or industrial progress, the disruptions of time attached to vertical imagery in *Extremely Loud and Incredibly Close* elicit a sense of the dissonance caused by seeing the Twin Towers collapse.

Extremely Loud and Incredibly Close's narrator, nine-year-old Oskar, has lost his father in the collapsing World Trade Center. The novel has drawn equal measure of castigation and praise; as Aimee Pozorski documents, *Extremely Loud* has been criticized for its perhaps "grating" narrator, and for the novel's ending, viewed by some quarters as "little more than a writerly farce." Despite uncharitable reviews, the novel's ending conducts a specific reverse, or a turn back; as Pozorski notes, the ending overturns "not only history but the laws of physics as well"; falling becomes flying in "a reversal of gravitational forces."[19] The ending is also in itself doubled. First, the narrative offers a textual description of Oskar locating pictures of a body falling from the Twin Towers:

> I reversed the order, so the last one was first, and the first was last. When I flipped through them, it looked like the man was floating up through the

---

19. Pozorski, *Falling after 9/11*, 125; 132; 126. Other discussions of the book's ending include Frost, "Still Life"; Lurie, "Falling Persons and National Embodiment"; Versluys, *Out of the Blue*, 46–47; Codde, "Philomela Revisited," 251; Mauro, "Languishing of the Falling Man," 596–99; Vanderwees, "Photographs of Falling Bodies"; Gessen, "Horror Tour."

sky. And if I'd had more pictures, he would've flown through a window, back into the building, and the smoke would've poured into the hole that the plane was about to come out of.[20]

The culmination of this temporal reversal is Oskar imagining his father alive, "safe" (326). Then the novel ends for a second time, now with a visual enactment of what Oskar has described (327–53). In the photographs that close the book, the reversal of time and gravity produce the sense that the impact of vertical trauma has been alleviated, or perhaps transformed. It is not irrelevant that *reverse* shares the roots of *vertical* I discussed earlier; the very concept of reversal as a turning back intimates the sense of a turning point repetitively enacted by the invocation of verticality.

Extremely Loud's semantic, and graphic, inversion of vertical descent is also prefaced by earlier moments in which the power afforded to falling is displaced by an emphasis on suspension. In a recalled tableau that resurrects the protagonist's dead father, Thomas, a bedtime story creates an imaginary "Sixth Borough" of New York, "separated from Manhattan by a thin body of water," and depicts a man jumping across the space, in so doing, allowing "every New Yorker" to feel they too are "capable of flight." Thomas then revises his description of "jumping," replacing it with the term "suspension" (217–18). The revision of words to depict verticality betrays the novel's overarching desire to countermand the distressing memory of those figures falling from the Twin Towers, individuals initially also referred to as "jumpers" before this phrase was replaced by more euphemistic terms that described victims having been "blown" or "forced" out of the buildings.[21]

While *Extremely Loud and Incredibly Close* exemplifies the combination of verticality and temporal disruption in its ending, the novel also betrays how certain affective modes are attached to vertical memorialization after 9/11. Oskar's discursive forays into how vertical engineering might be applied to both counterterrorism and memorialization are elaborate. What if, Oskar wonders, we were to build skyscrapers for the dead that extended down into the earth? The image fits into *Extremely*

---

20. Foer, *Extremely Loud and Incredibly Close*, 325. All subsequent citations come in the main text in parentheses.

21. For the controversy surrounding Richard Drew's image of the falling person, see Pozorski, *Falling after 9/11*, 11–14; 61–79; Mackay, "Representing 9/11," 13–15. For discussions of the term "jumpers" and its removal from media discourse after 9/11, see Junod, "Falling Man"; Pease, *New American Exceptionalism*, 153–79; Kaplan, "Homeland Insecurities."

Loud's repetitive return to ideas of verticality—and the horrors associated with the falling Twin Towers—after Oskar has been made bereft. His image not only echoes and reframes the concept of an afterlife as placed underground but also creates a mirror in which the dead are remembered by being physically located in a structure that is an inverse of the World Trade Center:

> So what about skyscrapers for dead people that were built down? They could be underneath the skyscrapers for living people that are built up. You could bury people one hundred floors down, and a whole dead world could be underneath the living one. (3)

Oskar's subterranean skyscraper is as much about transforming the vertical reach of the skyscraper into a structure that is invulnerable to collapse as it is about the creation of a physically descending graveyard. In this moment, *Extremely Loud* dramatizes the conversation between fictional texts and the cultural discourses that surround vertical events. It reveals the unique potency of vertical imagery after 9/11 while also vocalizing some of the same concerns articulated by critical debates surrounding the memorialization of the Ground Zero site.

The story of the Ground Zero site in New York is both the stimulus for this book's narrative and a microcosm of the verticality it considers. The place where the World Trade Center once stood continues to be determined by verticality. The site in Lower Manhattan is now occupied by two memorial pools, an expanse of carefully manicured trees, the 9/11 Memorial Museum, and, at the edge of the area, the previously termed "Freedom Tower," now rechristened as One World Trade Center. The structures that exist there are unequivocally enormous in multiple ways. One World Trade Center looms over the space; the memorial pools descend sixty feet; the museum contains an underground space that descends seven stories below the earth. In the years following 11 September 2001, the discourses that surrounded the process of rebuilding revealed a host of ideas attached to scale, time, and trauma. Marita Sturken speculated in 2001 that the more ostentatious designs for a new skyscraper or skyscrapers at the site would eventually be abandoned, but the finished memorial in fact sits next to One World Trade Center, now the tallest skyscraper in the northern hemisphere.

In the context of the debates surrounding what to rebuild at the site, the plans for One World Trade Center seemed to appease the kind of combativeness that was most starkly embodied by a proposal offered by several recovery workers at the Ground Zero site in the weeks after 11 September: to simply rebuild the Twin Towers one story higher but to include a "giant statue of a hand flipping off the terrorists" on the roof.[22] The new building, in fact, underwent several changes before finally involving a conservative plan, leading the architecture critic Paul Goldberger to ruminate that "we lost the opportunity to build a great building." Even further, the opportunity was also lost "to reassert American leadership." The new tower's form is to Goldberger not an adequate realization of the fact that "America is where the skyscraper began."[23] Building a higher structure than the World Trade Center recalls the commands made by the planners of the Chicago World's Fair to produce a structure that would "out-Eiffel Eiffel." It demonstrates continuing American competitiveness in the face of an ongoing global trend of constructing ever-taller skyscrapers, despite their lack of economic viability and the impracticality of their emergency evacuation. In the new building's simple form, height has become the primary means of articulating the idea of "American leadership" and the legacy of the skyscraper as a symbol of American innovation.

Goldberger's commentary is not unproblematic, conflating as it does principles and ambitions of architecture with the international standing of the United States. It also shores up the attachments to verticality that have recurred in the aftermath of 11 September 2001. The tower's vertical reach—perhaps showing the burden of too many meanings—has been subject to critique from other quarters. Sarah Senk suggests the building is defined by its "awkwardly tall aerial" that extends the tower's height to its goal of 1776 feet through a gesture that is "as forced architecturally as it is symbolically, recasting the events of September 11, 2001, in the mold of our Declaration of Independence from Britain."[24] The combined effect of these critiques is the revelation that the skyscraper—as previously the most recognizable, and now arguably also the most resonant, physical expression of verticality in American urban space—recurs as the site of competing discourses.

The memorial's pools plunge dramatically into the earth, making the exterior scale of the memorial as vast as One World Trade Center is, only along an opposing vertical axis. In an article titled "Ground Zero 9/11

---

22. Juliet McIntyre, a volunteer at the Ground Zero site after the attacks, offered this anecdote and is quoted in Nobel, *Sixteen Acres,* 19. See also ibid., 13–14, for the plethora of voices calling for a reconstruction of the Twin Towers; Simpson, *9/11.*
23. Goldberger, quoted in Pilkington, "9/11 Ten Years On," paragraph 24 of 29.
24. Senk, "Lost in Space," paragraph 5 of 11.

Memorial Flows with Mournful Splendour," Rowan Moore reported for *The Guardian* on 15 August 2011, "As each tower was big, each cascade is a cuboid Niagara, an inverted eruption, falling 30 feet to a flat basin, then another 30 feet through a smaller square hole." An "inverted eruption" is a strange and paradoxical choice of description that betrays the lingering effect of 9/11 in the somewhat violent noun "eruption"; among the sounds of the city, the cascades are in fact hardly audible as disruptive noise and rush over the sides of the basin with relative quietude. David Simpson's description of the site made its haunted-ness explicit when he noted the potential of the memorial to provide a disturbing echo of the events that occurred in the same place.[25] Among these readings of how the memorial is experienced, the visual contrast between the towering skyscraper and the descending pools compounds Marita Sturken's thoughts in "Memorializing Absence," an essay written soon after 11 September 2001. Sturken gestures toward the significance of verticality while not fully naming it; she argues that the discussions surrounding memorialization were furthering Michel de Certeau's "split view of the city" and "the contrast between the towering skyscrapers and the smaller acts of meaning created at street level." In this sense, Sturken notes, "The memory of this event already indicates the conflicting visions of the monumental and the individual, more intimate rituals of griefs."[26] Implicit is an understanding of the "monumental" and the "individual" that takes its power from a visual dialectic of "towering skyscrapers" and "smaller acts of meaning": a dialectic that is part of what this book calls the age of verticality.

As Marita Sturken notes, the 9/11 memorial and museum are sites that are "haunted by the stories of bodies."[27] Their "haunted" materiality is, further, bound up in ideas of scale and compression. The interior of the 9/11 Memorial Museum descends seven stories below the earth, comprising 110,000 square feet underground (477). The site is made of "bigness": the "vastness of the memorial pools," the "huge space of the underground museum," and the "immense size of many of the objects on display" (478). At the same time, the space also contains such enormous artifacts as The Last Column, a thirty-six-foot "colossus" made of steel that dominates the Foundation Hall, itself seven stories underground (471). The

---

25. Simpson, "Shortcuts," 22.
26. Sturken, "Memorializing Absence," paragraph 4 of 15; see also de Certeau, *Practice of Everyday Life*, 91–110.
27. Sturken, "9/11 Memorial Museum," 484. All subsequent citations come in the main text in parentheses.

FIGURE 1. The South Pool of the 9/11 Memorial, New York. Photograph by the author, September 2011.

Last Column is a "symbol of resilience, *because it survived*"; an "object of affirmation that mediates the loss and vulnerability experienced in the events of 9/11" (471, emphasis in original). And the museum's most controversial exhibit is referred to as the "composite." It is a block of compressed debris that was once four to five floors of one of the towers and is now only a few feet tall (484). To Sturken, these multiple expansions and compressions of scale "convey the sense of 9/11 as an event of massive importance"; the effect is that "scale functions in the site as a form of 9/11 exceptionalism, which itself forms a part of American exceptionalism" (478). The opposing representations of compression and vastness Sturken describes at the Ground Zero site are exemplary of the ways in which the memorial and museum produce frictional discourses of scale. The memorial comprises a miniaturized inversion of the buildings that once stood at the same site, a vast underground space within its museum, and the construction of a towering skyscraper at its edge. These facts do not merely reveal the competing desires that surround the question of how to remember 9/11. Rather, the inescapably vertical manifestations of the memorial indicate a much more complex process in which post-9/11 texts—not only fictional or even documentary narratives but also the texts that arise when a community memorializes an event and a national debate occurs about that memorialization—negotiate the resolutely vertical events that occurred on 11 September 2001.

As Sturken reasoned soon after 9/11, the space where the Twin Towers stood is "inescapably a graveyard." She proposed that, in spite of the various proposals for a new and conspicuous skyscraper to be built at the site, the final memorial would most likely focus "not on replacing the skyline but on rendering present the individuals who died there."[28] The memorial does, indeed, focus on the individuals who died, emphasizing their names in engravings on the structure. In the underground space of the museum and the two pools standing in the footprint of the towers and dropping a total of sixty feet down, the Ground Zero site is also an uncanny echo of Oskar's inverted skyscraper built for the dead. And Oskar's thoughts have been realized in other, perhaps unexpected, ways; the museum also contains, in Sturken's words, "an enclosed and restricted room that contains the remains of the dead who have not been identified."[29] Unsurprisingly, this fact has been the source of considerable controversy; it is a painful reminder that the events of 2001 produced a very

---

28. Idem, "Memorializing Absence," paragraph 8 of 15.
29. Idem, "9/11 Memorial Museum," 484.

particular kind of material absence that frustrates the course of understanding and memorialization.

Sarah Senk notes that "the focus on absence registers the disorder and bewilderment initiated by an event whose meaning might never be clear, even in the fullness of time."[30] In "The 9/11 Memorial Museum," Sturken, too, suggests that absence determines the 9/11 museum's strategies of presentation. Precisely because what happened on 11 September 2001 was an "extreme" event of disintegration and destruction, questions of memorialization and curatorship must ask: "How to fathom this kind of material transformation, the shock that these buildings, those desks and chairs, those bodies, those airplanes, were just gone?" To Sturken, this "incomprehensibility" ensures that, while the museum can be "deeply affecting" in narrating the "survival, resilience, sacrifice, and compassion" that occurred because of the events of 11 September 2001, it is "significantly less effective in addressing the *meaning* of 9/11" (484–85, emphasis in original). And how could it be any other way? The event eradicated the material verticality of two enormous towers, as well as the human life they contained. Those who died on 11 September 2001 are, in Jenny Edkins's words, "*ontically* 'missing'"; they have been removed from "a context in which they are part of their recognized social or symbolic system."[31] The *absence* of what was lost in its visually recognizable form (the bodies, the buildings, the ephemera of office life) haunts the site, making any attempt to understand the "meaning" of 9/11 already impeded.

Rather than trying to decipher any singular or cohesive meaning that might come from the events of 2001, this book advances from the critical examination of 9/11 and its associated texts in this introduction to make two claims. First, within the genre of post-9/11 narrative, representations of anticipative trepidation or temporal uncertainty cluster around representations of verticality. Second, studying verticality in post-9/11 texts uncovers the routes of affective modes of anxiety and uncertainty that recur across the twentieth century. The following chapters consider how 9/11 studies can be enriched by considering the emergence of verticality, both as a physical object and a metaphorical presence. The history of the Ground Zero rebuilding project in New York exemplifies how verticality infuses the texts that have come after 9/11. *Extremely Loud and Incredibly Close*'s representation of an inverted skyscraper for the dead distills

---

30. Senk, "Lost in Space," paragraph 11 of 11.
31. Edkins, "Time, Personhood, Politics," 129 (emphasis in original).

the ways in which various literary texts use the image of physical architecture not simply to represent a proprioceptive experience but rather to organize a certain set of ideas and discourses. These two texts, one physical (the memorial) and one literary (Foer's novel), speak back to a history of textual convergences between material verticality and its multiple nodes of reference.

A THEORY OF VERTICALITY: TRAUMATIC REPETITION, WORLDS OF FALLING, AND THE OVERWHELMING SUBLIME

If the textual representation of verticality in Foer's novel and the acts of rebuilding at the Ground Zero site engage with the specific vertical events of 2001, the work provided by 9/11 studies, the insights of trauma theory, and the critical paradigm of the sublime produce a theoretical entrance into the concepts discussed throughout this book. The vertical vocabulary of 9/11 suffuses the visual and textual realm in a way that echoes the repetitive structures of trauma theory, posing the question: how do we understand the representations that recur, the "images that never quite seem to capture the horror of the day, but yet emerge, as though via traumatic repetition, again and again and again?"[32] Verticality, then, is a mode of representation that grapples with anxiety and uncertainty best understood from a position that critically examines 9/11 and its narratives. This assertion provides the impetus and the methodological basis for the chapters that follow.

This book extends the work of scholars such as Aimee Pozorski, Marita Sturken, and Georgiana Banita, all of whom engage with the full complexity of claiming a post-9/11 canon while retaining specific focus on primary texts.[33] As Banita writes, the "appearances and disappearances of the terrorist attacks" in narrative necessitate a critical terrain that is as concerned with the oblique, elusive, and spectral presence of 9/11 as it is with the overt representation of the events of 2001 within fiction.[34] Pozorski goes further, to claim that a "crisis point in representation" has occurred as texts "seem to refer in excess" to the events of that day.[35] To Pozorski, the appearances and disappearances that Banita notes are

---

32. Pozorski, *Falling after 9/11*, 10.
33. Sturken, *Tourists of History*, "Memorializing Absence," and "9/11 Memorial Museum"; Pozorski, *Falling after 9/11*; Banita, *Plotting Justice*.
34. Banita, *Plotting Justice*, 2.
35. Pozorski, *Falling after 9/11*, 6–7; 1.

incontestable; they have produced a body of texts with which scholars must work to understand the problems of referring to the events of 9/11. Indeed, scholarship has reached a consensus that multiple narrative responses to 9/11 occur "aslant," to use Richard Gray's term; in the wake of the traumatic, "perhaps the way to tell a story that cannot be told is to tell it aslant, to approach it by circuitous means, almost by stealth."[36] At the same time, the host of meanings attached to the term 9/11 creates a semantic artery that is saturated with potential referents; 9/11 is now "semantically surrounded by an almost infinite list of peripheral terms, events, and ideas."[37]

The complex signification scholarship faces after 9/11 is evident even in apparently bare language. The terms 9/11 and *Ground Zero* are already themselves fraught; as Sturken notes, "Ground Zero" is "a name pulled from history"; the term was used to refer to the detonation point of the first nuclear bombs. The recirculation of the term to refer specifically to the location of the 2001 attacks in Lower Manhattan both occludes the other crashes that occurred (American Airlines 77 at the Pentagon and Flight 93 in Pennsylvania) and also cements a narrative of 9/11 as "a moment in which the United States lost its innocence."[38] The term 9/11 is no less problematic for its condensation of multiple reference points into a single atemporal figure.[39] The language of 9/11 exemplifies the broader risks associated with marking 9/11 as an exceptional event beyond the scheme of ordinary understanding or, as David Simpson puts it, "an interruption of the deep rhythms of cultural time, a cataclysm simply erasing what was there rather than evolving from anything already in place."[40]

This burdened context of talking about 9/11 produces a double risk. The omnipresence of 9/11's visual vocabulary means that its effects ripple out in narratives too broad to claim as a discrete field. And this very same hemorrhage also ensures that talking about the representation of 9/11 can secure the exceptionalism that scholars have been at pains to reject: by noting what Pozorski calls an "excess" of representation, we risk complicity with a reductive tendency that marks 9/11 as a moment that

---

36. Gray, "Open Doors, Closed Minds," 136.
37. Bragard, Dony, and Rosenberg, introduction to *Portraying 9/11*, 3.
38. Sturken, *Tourists of History*, 167.
39. I use the term 9/11 throughout this book to refer to the events of 11 September 2001 and their aftermath, since it is the most familiar way to do so, while recognizing that inherent problem of using a term that might signify any or every 11 September and not merely that of 2001. This point is usefully discussed by Jacques Derrida in Borradori, *Philosophy in a Time of Terror*, 85–87.
40. Simpson, *9/11*, 4.

changed everything. In 2006 Simpson proposed that, with time, "it may come to appear that 9/11 did not blow away our past in an eruption of the unimaginable" but in fact that "it refigured that past into patterns open to being made into new and often dangerous forms of sense."[41] *Waiting for the Sky to Fall* proceeds from the understanding that the emergence of a post-9/11 genre—if it is possible to claim such a porous entity—is bound up in the history of verticality. As Banita puts it, post-9/11 narrative "testifies to a new set of anxieties about how to relate the present to the past, but also about how knowledge of the past (and its residual traces) inflects our understanding of the present, seen not as a break with history but as its organic outcome."[42] Further, as I discuss in the following, the temporal circuitry through which trauma is conveyed is the perfect analogy for the process of re-understanding verticality after the fact of 2001.

The historical perspective predicted by David Simpson is essential to this book's claims about verticality as a figure that cements the longer patterns of talking about anxiety, time, and affect across the twentieth century. Throughout its course, *Waiting for the Sky to Fall* strives to close the gap between 9/11 understood as a necessarily problematic critical category and 9/11 as a moment that allows a better understanding of narrative representation across the long twentieth century. In order to do this work, it is helpful to consider certain gaps in scholarship. There have been critical interventions in the field of 9/11 studies devoting attention to the trope of falling, but as yet no other study considers verticality as a trope in itself.[43] This absence is in part explained by the problem of claiming verticality as a discrete metaphor. While the genre of 9/11 narrative is rich for its complexity, its patterns of diffusion, and its porousness, verticality also crosses formal and generic boundaries, encompassing metaphors of flight and falling, and referring to both height and depth. I work to circumscribe the parameters of verticality in the following by exploring critical theory's legacies of vertical representation, beginning with how the reiteration of "the man falling" illustrates what deconstruction and trauma theory call "the problem of reference."[44]

---

41. Ibid., 13.

42. Banita, *Plotting Justice*, 4.

43. Pozorski offers a comprehensive study of the existing field in *Falling after 9/11*. Her own work is indispensable to an understanding of the post-9/11 figure of falling; as she notes, "Of those critical works that take up the 'problem' of 9/11, none of them offers a sustained look at the figure of the falling man—at once the most-cited and perplexing image to emerge in our national consciousness from that day" (7).

44. Pozorski, *Falling after 9/11*, 10.

Cathy Caruth's seminal account of falling in the 1996 volume *Unclaimed Experience* prompted a dense cluster of scholarship that unpacks gravity as a narrative trope, and established the theoretical impetus of verticality as bound by the facts of gravity.[45] Caruth begins her study by outlining the anxieties leveled against deconstruction: namely, that poststructuralist interventions into literary studies might "amount to a claim that language cannot refer adequately to the world and indeed may not truly refer to anything at all." She continues by exploring Paul de Man's response to this anxiety (in the 1982 essay "The Resistance to Theory"), suggesting that de Man's repeated use of a story of falling to explain his argument points toward the figure of falling itself as a method of understanding reference. For de Man, seventeenth-century epistemology provides an exemplar of how "the problem of reference became, in the history of thought, inextricably bound up with the fact of literal falling." This leads Caruth to the following statement:

> The world of simple motion was ended, once and for all, with the discovery, by Newton, of gravitational force, or the revolutionary notion, introduced in Newton's *Principia*, that objects fall toward each other.... It could be said, indeed, that with this assertion, the world of motion became, quite literally, a world of falling.[46]

In Caruth's reading of de Man, and in Pozorski's gloss of these texts, falling "is a figure for the difficulty of referring to something that is impossible to fully assimilate into consciousness and the system of language."[47] As Pozorski points out, Caruth's insight is that "Newton's theory reveals that everything is falling all of the time, so there is never a stable reference point." This leads to a startling realization: "What we call 'falling,' then, is just a special case of something that is always true." Pozorski uncovers the significance of falling that this book takes up in its discussion of verticality when she suggests that "part of the problem for Caruth is that we think there is solid ground somewhere, but in fact we are falling all of the time; further, everything is falling—even the thing we are falling toward."[48]

---

45. Caruth, *Unclaimed Experience*, 73–90. See also Pozorski, *Falling after 9/11*, 25–39; Kaufman, "Falling from the Sky."
46. Caruth, *Unclaimed Experience*, 73–74; 74; 75–76.
47. Pozorski, *Falling after 9/11*, 10.
48. Ibid., 11; 32.

Caruth is concerned with "not simply confronting the science behind falling but confronting the fact that there are, perhaps, no words to refer to such a moment." Whereas Paul de Man and Cathy Caruth after him interpret falling as a figure of the "unimaginable," post-9/11 writers "have the additional problem (and responsibility) to represent the falling man as a particular and literal truth of 9/11."[49] Pozorski argues that, in a post-9/11 context, narrative representations of falling produce a surfeit of words and images: a different, but no less significant, crisis for representation. This book builds on the claims made by Caruth and Pozorski to suggest that a world of falling—an existence in which everything is always already falling—is a helpful way to understand the uncertainty and disorientation that emerges from verticality. The realization that everything is falling, as Pozorski puts it, removes a "stable reference point" and signifies the disruptions that occur in representations of verticality. Further, in the production of temporal uncertainty, post-9/11 texts reveal the ways that vertical imagery fuses with concepts of time necessarily informed by trauma theory.

My reading of Foer's novel exemplifies how the representation of traumatic temporality in post-9/11 narrative sutures the vertical experiences of 11 September 2001 with a genealogy of literary, visual, figurative, and physical verticality. *Extremely Loud* marshals vertical imagery to grapple with the events of 11 September 2001 while also soliciting textual occasions that arrest a fall or represent a vertical collapse made static, frozen in time. This juncture between gravity and time in Foer's novel exposes the significance of vertical representation in relation to trauma theory. The concept of trauma itself has evolved into a fluid category characterized by the confluence of bodily, psychic, cultural, symbolic, historical, and structural currents; trauma's expression as "a point of intersection, of turbulence," secures its "powerful force."[50] As Pozorski notes, trauma theory offers an essential method of studying post-9/11 narrative, but hers is one of the first to apply the rigors of that field to texts that have come after the events of 11 September 2001. This book extends some of the critical concerns put forward by Pozorski but stands with Michael Rothberg's formulation of trauma theory as *"necessary but not sufficient for diagnosing the problems that concern us as scholars and human beings."*[51] This book uses the imperatives of trauma theory to explore the

---

49. Ibid., 11; 25.
50. Buelens, Durrant, and Eaglestone, introduction to *Future of Trauma Theory*, 1.
51. Rothberg, preface to *Future of Trauma Theory*, xiii–xiv (emphasis in original).

relationship between 9/11 and verticality, while emphasizing the value of other critical theory from fields such as cinema, philosophy, and history. The age of verticality traced in the following chapters is one that repeatedly imagines time-stopping moments of ruin, spectacular explosions, and terrible falling. These moments are indiscriminate in the shape they take; verticality moves over genres of prose, poetry, documentary, photography, and cinema, and offers itself in texts drawn by hand, witnessed firsthand, and imagined in extensive metaphor. Trauma theory is therefore a helpful node of examination for verticality, especially with its relationships to time and 9/11; trauma theory alone is also by itself insufficient to the task of understanding the complex generic and formal qualities of the texts discussed in this book.

Reflecting on the future of trauma theory, Rothberg notes that, "however we conceive it," trauma is "a category that ought to trouble the historicist gesture of much contemporary criticism as well as its concomitant notions of history and culture." One legacy of theorists such as Caruth is the claim that "trauma dislocates history." Nonetheless, acknowledging the "limits of classical trauma theory's dislocation of its own context of emergence (i.e. its failure to transcend a Eurocentric frame)" does not make it easier to imagine trauma free from a concept of dislocation— of "subjects, histories, and cultures."[52] Nowhere is this more evident, perhaps, than in trauma's relationship to time, a relationship that is most helpful to an understanding of verticality. Caruth glosses the cause of trauma in Freud's work as "an encounter that is not directly perceived as a threat to the life of an organism but that occurs, rather, as a break in the mind's experience of time."[53] Put another way, traumas are, by definition, the events that cannot be assimilated into "linear narratives"; trauma is not felt in "homogenous linear time"; "there are no words, no language, through which such an experience could take place." The delayed response to the trauma and its perception after the fact means that traumatic events "are only experienced, if we can call it that, when the past, which has not yet 'taken place,' intrudes into the present and demands attention."[54] If trauma is that which imposes belatedness upon the subject, post-9/11 narratives such as *Extremely Loud* attach the temporal fluctuations of trauma's power to specifically disruptive vertical metaphors. Vertical imagery in *Extremely Loud* that expresses a reversed

---

52. Ibid., xii.
53. Caruth, *Literature in the Ashes of History*, 5.
54. Edkins, "Time, Personhood, Politics," 132–33.

descent or the holding still of a fall comes, firstly, via a fusion of visual and written media, and, secondly, through a reversal of narrative chronology. The novel makes vertical imagery a remedy for anxieties specifically induced by those events haunting its margins; here, vertical imagery reveals a desire to work through the anxiety generated by the visceral experience of seeing the Twin Towers collapse.

Extremely Loud and the rebuilding of the Ground Zero site raise questions about how to memorialize an overwhelmingly vertical, spectacular, and traumatic moment, but they also gesture toward a fundamental way of conceiving other vertical events. For the occasions that often provide the inspiration for the texts discussed in this book, a key theoretical paradigm is that of the sublime, which is intimately related to the representation of overwhelming height or vertical disintegration. Edmund Burke's formulation of that which stimulates a "passion" in the beholder "in which all [the soul's] motions are suspended" reflects the tendency among eighteenth-century philosophers to announce the sublime as a moment entangling pleasure with terror.[55] As Christine Battersby remarks, in Burke's scheme, "whatever is visually terrible is also necessarily sublime."[56] At the end of the eighteenth century, Immanuel Kant depicted the sublime as the confrontation between the "I" and the element that has the potential to completely destroy it.[57] Nineteenth- and twentieth-century theorists tended to emphasize the "failure" of "understanding and reason" to assimilate the sense of infinity invoked by the sublime. Throughout this history, terror is an essential part of the way the sublime has been conceptualized. Burke's initial argument—and the one most commonly associated with his name—was that terror is a necessary part of the sublime; on the other hand, Kant distinguished terror in the sublime from real fear.[58]

Despite these oscillations, from the late seventeenth century onward, the varying set of conditions appended to the sublime have included two recurring ideas. The first is perhaps the most famous and describes the entangled enjoyment and fear of the sublime moment: what Battersby calls the "pleasurable shudder." The second is that the pleasure of the sublime experience is generated by the confrontation with "an object or an entity" that is felt by the beholder to exceed the limits of understanding and, as they are faced with the "overwhelming," "awe-inspiring," and

---

55. Burke, *Philosophical Enquiry*, 57; Battersby, *Sublime*, 1.
56. Battersby, *Sublime*, 25.
57. Ibid., 1.
58. Ibid., 1; 24–30. See also Kant, *Critique of Judgement*.

"unrepresentable," they are at least temporarily barred from fully comprehending what they are seeing.[59] This book is concerned with the marriage of visual spectacle with temporality in narratives where verticality is as revelatory as it is disorienting. Throughout the course of the twentieth century, the representation of vertical spectacle in American narrative marks a variety of sublime experiences. This book examines how these representations draw out the inherent paradoxes of the sublime: as an experience that through its very mediation risks becoming banalized and yet represents the opposite of banality; as an experience that has come to mean a response to not merely the natural world, works of art, or political acts, but also the man-made shapes of the city; as a concept essentially describing a moment that exceeds understanding and that also must be somehow assimilated into human consciousness; and as a moment that can prompt pleasure in witnessing disaster but equally generate questions about moral culpability.

THE AGE OF VERTICALITY

Waiting for the Sky to Fall begins with the specifically dissonant visual effect of the skyscraper in texts that represent Manhattan space, focusing on the graphic work of Winsor McCay, in which ideas about the structure of the human mind convene with the symbolism offered by concrete vertical structures and events. Chapter 1, "'Down Down Down He Shot': Winsor McCay's Work, Vertical Collapse, and Time in the Modernist City," considers McCay's long-neglected comic strips and animated films from 1904 to 1917, focusing on *Little Nemo in Slumberland* and *Dream of the Rarebit Fiend*, which propose a fantastical world of dreaming determined by repetitively vertical occasions: oversized skyscrapers, elevated and often collapsing bridges, or the flight or fall of its protagonist. Chapter 1 unfurls from a recognition of the prominence of a new fascination with what Virginia Woolf would call the "dark places of psychology."[60] Essential to this chapter is Sigmund Freud's understanding of the mind, which describes the psyche as a cumulative effect of the interplay between the conscious and unconscious that is conducted by and infused with psychic energies, and created the foundation of his dream theory.[61] The

---

59. Battersby, *Sublime*, 1.
60. Woolf, "Modern Fiction," 192.
61. Freud, "Third Lecture," reproduced in Rosenzweig, *Freud, Jung and Hall*, 420; see also 409.

Freudian description of a psychic topology, in which the mind is a kind of intricate building, is the basis for chapter 1's consideration of both how McCay's comic strips are engaged in a similar process of vertical construction, as well as the goal of unraveling the ways in which the mind works.

While chapter 1 poses New York's spatiality as one creating a sense of urgency and claustrophobia—a feeling that there is literally nowhere else to go but up—chapter 2 extends this argument into the vertical activities of mining and oil drilling that transduce energies of expansion no longer fulfilled by westward expansion. In chapter 2, "Upton Sinclair's Vertical Infernos: Oil Procurement and Disaster Culture," these ideas are framed in terms of how the new architectural forms of skyscraper and oil derrick might be seen to work against a particularly American condition of spatial transience; responding to the end of apparently limitless horizontal space with the less circumscribed space of the sky or earth. Chapter 2 ties the effects of the frontier's closure to the new methods of advancing across horizontal space at great speed via train and automobile in Upton Sinclair's *Oil!* (1927) and examines the novel's emphasis on the vertical action of oil drilling. The discussion considers how *Oil!*'s conflicting tones simultaneously extol, and caution against, the economic success ensured by the oil industry and examines the consequences of this double pull on its representation of the oil geyser and oil rig fire.

Chapter 3, "'The Horizon Was an Illusion': Flight, Escape, and Imagining Vertical Space in Leslie Marmon Silko's *Almanac of the Dead*," considers verticality in Silko's book and Native American writing more generally, foregrounding the significance of the expulsion of indigenous people from spaces as white expansion occurred on the North American continent. Governmental cartography has defined the 400-year-long history of contact between Native American and Euro-American people. Immediately after the American Revolution, Congress proceeded to legalize both the persecution, and the attempted assimilation, of Native American people, primarily through legislature that effectively changed the way that the continent was mapped. Under Andrew Jackson's presidency, Congress passed the Indian Removal Bill in 1830, legalizing the annexation of Native American territory lying east of the Mississippi; those tribes previously living in these regions were forcibly removed to what is now Oklahoma.[62] By 1890, Native Americans had been squeezed into

---

62. Porter, "Historical and Cultural Contexts," 63ff. See Kessler and Jacobs, *Mapping America*, 112–15, for maps showing Native American territory, federal lands, and reservations in the United States.

comparatively small spaces such as those in the state of Washington. President Jackson's attempt to give the Indian Removal Bill a "philanthropic gloss" could not hide the deeply rooted desire for the Native American population to simply "disappear" in order to fulfill the dictates of Manifest Destiny and expansion to the Pacific.[63]

These energies of expulsion via horizontal movement across the map were unable to contain the limits of the space under pressure, and they inevitably stalled, erupting in bloody clashes between Native and Euro-Americans. Native American texts frequently depict the damage inflicted by westward expansion in narrative gestures that transpose the permanency of that history into an imagined vertical form, one which both summons up the historical reality of the frontier and also frequently expresses hope for a better future for indigenous people. Imagining verticality, then, is an escape from the constricted movement produced by myriad federal policies such as those of the 1850s that confined Native Americans to small areas in the hope of curtailing intercultural contact. Gerald Vizenor's "survivance" describes the ways in which Native American texts exhibit a sustained desire to ensure survival of cultural traditions, referring to narrative modes adopted by Native American texts to confound dominant or subjugating cultural traditions. Vizenor's theory of the "ironic crease" of what he calls "postindian" textuality is based upon Native American trickster discourse. Survivance suggests that literary and cultural texts born from a history of subjugation can find oblique ways to renarrate, and therefore emphasize anew, a resistance to the constraint set by Euro-American discourses.[64] Chapter 3 engages with these critical paradigms to argue that verticality is frequently the vehicle for an expressed desire to ensure the endurance of cultural and literary traditions.

The new technology that facilitated the vertical spectacles of the skyscraper, the oil geyser, the airplane, and the dynamo in the first quarter of the twentieth century met with a reprisal at the midcentury when science endeavored to send man into space and to drop calamitous bombs from above. Chapter 4, "Terror/Power: Allen Ginsberg's Nuclear Poetics and the Space Race," considers how Allen Ginsberg's poetry and prose that engage with new nuclear technology—both that which facilitates the atomic bomb and launches space crafts—have been overlooked in favor of his more famous work. The chapter asks how Ginsberg's work mobilizes vertical imagery to refute a straightforward response to the nuclear

---

63. Porter, "Historical and Cultural Contexts," 51.
64. Vizenor, *Fugitive Poses*; idem, *Narrative Chance*.

that is either wholeheartedly condemning or unblinkingly supportive. It considers how the inherent performativity of poetry informs Ginsberg's protests against political and cultural acts that are directed specifically at the nuclear industry and Vietnam War. The chapter marks a renewed understanding of the "sublime" as a way of working with the effects of scientific and human change.

Chapter 5, "Traversing Vertical Space: Philippe Petit's 1974 Wire-Walk, Danger, and Transformation," focuses on a moment of crossing over vertical space: Petit's wire-walk between the Twin Towers in 1974, which I discussed at the beginning of this introduction. To John Carroll, writing soon after the events of 11 September 2001, the World Trade Center was both a symbol of Western hubris and an invitation to terrorism. He argued that it was precisely the beauty of the Twin Towers—specifically their "sheer vertical precision"—that invited their destruction. The beauty of the structures, to Carroll, was a "warning"; it was the "excess" signaled by their literal height as well as their symbolic resonance that inspired the events causing their destruction.[65] Reading James Marsh's film *Man on Wire* as well as comics and prose that represent the wire-walk, chapter 5 interrogates the ideas attached to the World Trade Center and how they participate in a longer history of anxiety surrounding the skyscraper. It reprises the earlier discussions of the sublime surrounding vertical events and movements, revealing the significance of the sublime experience in post-9/11 narrative. It argues that *Man on Wire* is distinct for the way in which it invokes what would typically be called a sublime experience as one that enacts a chronological retreat, crucially as a means to unravel the disturbing associations of verticality left by the events of 2001.

The coda to this book, "Up and Down Stories," provides a close reading of Chris Ware's collection of overlapping graphic texts and comics, *Building Stories* (2012). The coda condenses the emphasis in the previous chapters on multigeneric texts, operating as it does at the juncture of various kinds of narrative form. *Building Stories* exemplifies a way of reading verticality via figures of constructing meaning and history, and through its most explicit metaphor, the visual representation of vertical Chicago buildings "as" human stories. The collection comprises several separate texts that can be unfolded like blueprints or made into vertical cardboard buildings. By making its prime metaphor the physical reality of the graphic art contained within the collection, *Building Stories* exposes

---

65. Carroll, *Terror*, 15–26.

representative methods essential to understanding verticality across the long twentieth century. Crucially, Ware's work employs a nonsequential form in conjunction with vertical imagery to represent the personal histories of its protagonists as ones that exist in a fluid chronological order. Throughout its course, this book unveils distortions or manipulations of temporality that come via a text's formal strategies and that solidify a sustained marriage between form and function in texts emphasizing discussions of time as a complement to vertical imagery. By problematizing a straightforward sense of time's progression, Ware's collection reflects upon an idea that emerges as central in this book: that verticality is profoundly connected to a sense of waiting for something that may or may not happen.

CHAPTER 1

"DOWN DOWN DOWN HE SHOT"

Winsor McCay's Work, Vertical Collapse, and Time in the Modernist City

WINSOR McCAY'S DREAM-COMICS AND MODERNISM

In 1907, Henry Adams yoked the new psychological models associated with dreaming to vertical metaphor. In the revelation of a "sub-conscious chaos below," he imagines the individual as "an acrobat, with a dwarf on his back, crossing a chasm on a slack-rope, and commonly breaking his neck."[1] Adams's first metaphor makes an imagined chasm in the mind, a "chaos below" that communicates an anxiety about the concept of the unknown posed by new psychological claims of consciousness. His second metaphor advances his first, making an awareness of one's own consciousness into a precarious journey over a vertical "chasm." This doubled vertical representation prefigures the association of the tightrope walk with uncertainty discussed at length in chapter 5 of this book. Underwritten by anxiety, Adams's metaphors also gesture toward a principal psychotemporal modality that emerges at the beginning of, and extends through, the twentieth century.

---

1. Adams, *Education*, 362.

Adams's prose has particular affinity with the sustained use of verticality in the early American comic strips by Winsor McCay that are the basis of this chapter's discussions. The texts I refer to as the *dream-comics*—McCay's *Dream of the Rarebit Fiend* (published from 1904 to 1911 and during 1913) and *Little Nemo in Slumberland* (published from 1905 to 1914)—explore the chimerical world of sleeping, depict the disorienting experience held by dreams, and examine how dreaming might be understood.[2] Both series provide the complex negotiation of a dream-life filtered through the primary metaphor of various vertical events. *Rarebit Fiend* was a popular, and purposefully adult, black-and-white comic strip with a single dream narrated in each installment. *Little Nemo,* one of the first American comic strips to be published in full color, echoed the same structure. In the first panel of each *Little Nemo* strip, a boy called Nemo falls asleep and dreams he is summoned to meet Morpheus, the King of Dreams. Borrowing from classical mythology in its use of the Greek god Morpheus, *Little Nemo* proposes fantastic, vertically driven journeys that outsize skyscrapers or magnify human bodies, elevate or collapse bridges, turn worlds entirely upside down, and result in the flight or fall of its protagonist.

Throughout its course, McCay's work dialogues with contemporary discourses surrounding time. Jürgen Habermas suggests that modernist culture, placing "new value" on "the transitory, the elusive and the ephemeral, the very celebration of dynamism," exposes "a longing for an undefiled, immaculate and stable present."[3] Representing a destabilized world that is physically falling apart, and at the same time suggesting a fracture of previous abstract epistemological models, *Rarebit Fiend* and *Little Nemo* expose precisely this relationship between the transitory and a desire to hold time still. The modernist impulse, famously characterized by Charles Baudelaire as a recognition of "the transient, the fleeting, the contingent,"[4] is overtly reflected by the formal conditions of *Rarebit Fiend* and *Little Nemo*; every strip of both series concludes with a panel depicting the dreamer waking. As such, the dynamic of McCay's dream-comics is a circular narrative of falling asleep/dreaming/waking, a repetitive excursion into a carnivalesque world that is always terminated by the

---

2. For a full discussion of the evolution of *Little Nemo* and *Rarebit Fiend*, see Roeder, *Wide Awake in Slumberland,* especially 7–9; 45–77; Canemaker, *Winsor McCay,* 78; 97. The *Little Nemo* series is published in Winsor McCay, *Little Nemo*. Throughout this chapter, I provide the original publication date of each strip.

3. Habermas, "Modernity," 1750.

4. Baudelaire, "Painter of Modern Life," 403.

final return to waking life. As this chapter discusses, *Little Nemo* reflects Habermas's depiction of a modernist infatuation with dynamism and also finds new ways to represent that dynamism.

The first American comic strips were appended to newspapers referred to, in aggregate, as *yellow journalism,* a phrase coined in 1897 at the height of a circulation war between William Randolph Hearst and Joseph Pulitzer that referred to the print color used for the protagonist of the very first comic, the immigrant urchin "Yellow Kid" in Richard F. Outcault's *Hogan's Alley* (published from 1895 to 1898). Yellow journalism was aimed at the masses and offered articles that were a blend of fact and fiction; comic strips created worlds that were an apt extension of the scandalous articles, pseudoscientific claims, and fabricated interviews within the newspapers they supplemented.[5] McCay's depiction of a fleeting dream-world was not greeted with the same enthusiasm as that for the more farcical and often violent story lines of *Hogan's Alley,* Rudolph Dirks and Harold H. Knerr's *The Katzenjammer Kids* (1897), and Frederick Burr Opper's *Happy Hooligan* (1900). The children at whom *Little Nemo* was directed were especially bewildered by the series' subject matter and its whimsical tenor; as the popularity of the "Yellow Kid" indicates, Hearst's and Pulitzer's working-class audiences were more attuned to the plight of a plucky, impoverished child than that of the vulnerable, middle-class Nemo.[6] And despite the work of scholars such as Katherine Roeder and Scott Bukatman, *Little Nemo* remains overlooked by contemporary scholarship, despite its acknowledged influence upon prominent graphic artists.

Among the innovations of early twentieth century narrative, the American comic strip is frequently crowded out of view or disregarded altogether in preference of literary and visual texts with more overt ties to an emerging modernist aesthetic. Yet McCay's work specifically refracts the way that early psychology and psychoanalysis disoriented existing ideas of consciousness and, as this chapter discusses, *Rarebit Fiend* and *Little Nemo* produce a complex intersection between the vertical disruption of their overt narratives and the sense of disruption enacted by their formal maneuvers. More formally and conceptually experimental than contemporary strips such as *Happy Hooligan, Little Nemo*'s interpretation of what a comic strip might look like and what story it might tell also

---

5. Mott, *American Journalism,* 538ff.; Tebbel, *Compact History of the American Newspaper.*
6. Blackbeard, "Greatest Strip That Ever Flopped," 5–7.

anticipates the innovations of later modernism. American comic strips are rarely discussed in relation to other forms with which they have affinities, such as Cubism and early modernist fiction. Yet comic strip form was uniquely suited to representing the pressures exerted by the ideas of time and space that were primary concerns of modernist discourse; in its structure, the comic strip spoke both to the spatial distortions of fine art and the antilinear narratives of other modernist texts, and the relationship between comics and these other narrative forms is crucial to an understanding of how work such as McCay's mediated the questions of representation that preoccupied artists of the period.

Thomas Vargish and Delo E. Mook suggest that a primary reason why modernism lacks a straightforward critical history is that it was resolutely "subdisciplinary or transdisciplinary in its operation" and, accordingly, to study it requires a multidisciplinary method. Vargish and Mook further suggest looking to three "cultural diagnostics" in order to untangle "the underlying values of the period," using the phrase to refer to "any human activity or production that may be analyzed in order to abstract the historically defining values." Their analysis focuses on three instances of epistemic and representative revolution: relativity, the theory which produced new understandings of space and time; Cubism, the form which transformed spatial representation in two-dimensional art; and modernist fiction, which disrupted and rebuilt the concept of temporality in literary texts. My methodology significantly differs in that it does not argue for a unified sense of what Vargish and Mook call "historically defining values": a phrase already fraught with certain slippages and assumptions. However, the three cultural diagnostics chosen by Vargish and Mook reveal the interpenetration of modernist media forms as well as the way it has been traced by scholarship.[7] Such scholarship reveals a lacuna that this chapter works to fill: namely, the relationship between the comic strip form and the epistemic and representative revolutions associated with modernism.

A variety of thinkers between 1870 and 1930 articulated a loss of faith in the idea that modern "ethical, political, and aesthetic ideals" were "destined to fuse with scientific, technological, and economic advances and lift humanity into a new life."[8] By consistently attaching vertical representations to notes of anxiety, McCay's work, too, suggests less an untroubled anticipation of modern progress and more a concern with

---

7. Vargish and Mook, *Inside Modernism*, 4; 6; 4.
8. Brenkman, "Freud the Modernist," 172.

the immediate impact of new ideas and technologies upon structures of knowledge. Scott Bukatman marks the kind of vertical distortions and transformations within McCay's comic strips as a ludic experimentation with the idea of plasticity, arguing that both *Rarebit Fiend* and *Little Nemo* expose the anxieties of early twentieth-century America through a sense of "*queasiness*" about the modern world.[9] Bukatman reads the metamorphosis of worlds contained within McCay's dream-comics as interruptions to the supposed orderliness and structure of the perceived world; he does not, however, tie the repetitive verticality of these texts to the disruptive nature of their form and content. There are few installments of *Little Nemo* and *Rarebit Fiend,* in fact, that do not feature some kind of significant vertical movement or distortion, whether in the motifs of flight and falling, collapsing buildings and upright structures, up becoming down, or what is short becoming excessively tall. Verticality in McCay's work is crucially related to the radical destabilization Bukatman describes, and yet it is a phenomenon that is so prevalent it has been all but ignored by previous scholarship.

A *Rarebit Fiend* strip from 1906 (figure 2) is typical of the way in which verticality emerges in McCay's work. The strip begins with the dreamer occupying the corner seat of a street car. Several oversized people proceed to sit next to him on the bench; as the dreamer becomes more squashed into the wall of the car, the vehicle begins to ascend a steep hill. The dreamer offers interjections of surprise and consternation as the car twists on an axis, until he is at the bottom of a pile of fellow travelers exclaiming, "Oh! Oh! Oh! I can't get out oh! I'm squeezed to a whisper. This is a bum street car route. I'll die if I don't get out of here. These fat people are smothering me, help! Help oh help! Oh oh oh oh." The six panels torque on an axis, twisting around so that the protagonist begins in an upright position and ends lying horizontally under a vertical pile of bodies. The fixed perspective in the strip produces "a smooth sequence of transitions," a quality that Bukatman suggests makes *Rarebit Fiend* McCay's "most 'cinematic' comic strip."[10] The technique also creates a visual approximation of gravity at work; the reader's perspective remains consistent while the street car becomes vertical. The installment is exemplary of the various currents that recur throughout *Rarebit Fiend* and, to some extent, *Little Nemo*. It is formally typical of the series, not merely in its use of fixed perspective, but in its highly structured narrative, which

---

9. Bukatman, *Poetics of Slumberland,* 51 (emphasis in original).
10. Ibid., 50.

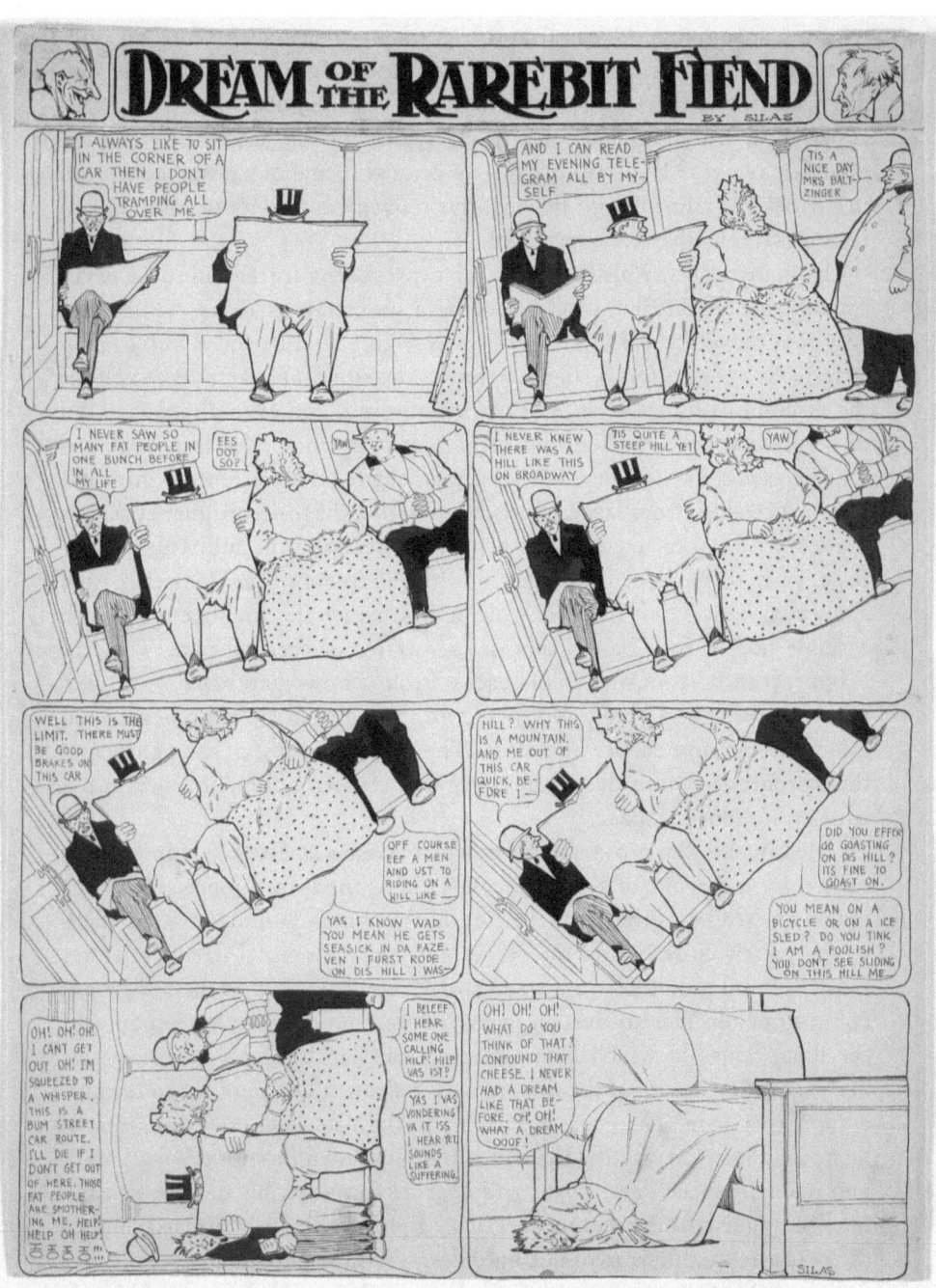

FIGURE 2. *Dream of the Rarebit Fiend*, Winsor McCay, 1906. Ink over graphite underdrawing with overlay. Original drawing held by the Library of Congress, Prints and Photographs Division, LC-DIG-ppmsca-07860.

initiates an apparently banal and everyday scenario (traveling on public transport), transforms that scenario into an increasingly strange and discomforting experience, climaxes in the frightening realization of that experience, and culminates in a panel typical of the series, in which the dreamer awakes. The strip provides a central image of vertical confusion that works in tandem with its narrative movement into a disorienting and uncomfortable physical experience.

In many ways, *Rarebit Fiend* set a precedent for the kind of vertical representation that would become a mainstay of *Little Nemo*. The very first strip of the *Little Nemo* series (figure 3), like the *Rarebit Fiend* strip from 1906, offers a single dream as the site of its narrative, explores the discomforting experience of vertical motion, and exemplifies the kind of formal experimentation typical of McCay's work. The frames cumulatively enact a visual narrative independent from their textual information, proceeding from large panels filled with light tones to smaller panels shaded with darker colors. The first panel occupies the full width of the page, only to be bisected precisely in the following four lines that each show two panels half the width of the first panel and finally ending with a single line of four panels that are a quarter of the first panel's width. From top to bottom, the colors move from yellows and ochers, through reds, to greens and blues, and finally an inky black, as Nemo is depicted falling deeper and deeper into sleep. The decreasing size of the panels and their progression into deeper hues evoke a claustrophobic journey into a smaller and darker space.

Meanwhile, the prose narrative depicts Nemo at first safely in his bed, visited by a servant of King Morpheus, and then on a journey into the realm of dreams riding upon a flying horse named Somnus. As the edges of the panels close in and the colors deepen, the textual narrative states that "Somnus stumbled on a star" and "Nemo clutched at the saddle but could not hold fast so over he went." After the comparable expanse of the first frames in which Nemo lies statically in bed, the shrinking size and darkening colors of the final panels indicate a disorienting and frightening experience. The prose abets this in both its semantics and its syntactic arrangement, stating that "down down down he shot through miles and miles of space" and "over and over he turned in his descent causing intense anguish and becoming so dizzy that he thought he was going to die." The textual narrative not only describes a frightening experience of falling through limitless space but also approximates the seeming inevitability of this fall by omitting conjunctions or punctuation. Nemo is not merely tumbling; he is falling unchecked. The size and color of the panels

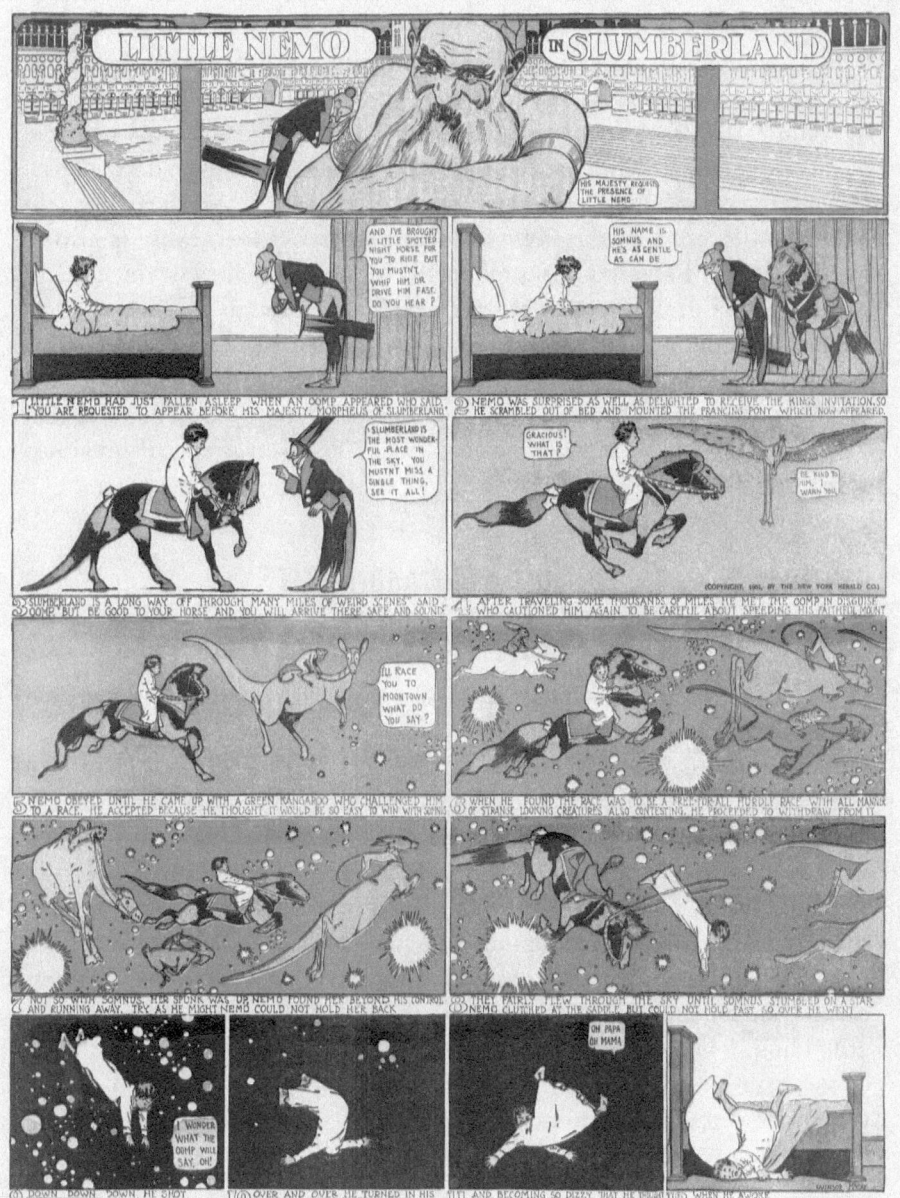

FIGURE 3. The first of Winsor McCay's *Little Nemo in Slumberland* series, published in the *New York Herald*, 15 October 1905. Courtesy of Peter Maresca, Sunday Press.

that depict Nemo's fall visually parallel the fear he experiences, while the prose both describes this fear and also augments it with an uncontrolled cadence.

The verticality exemplified by these strips is a visible surface concept: a graphic rendering of worlds turned upside down, tall buildings collapsing, and falling that defines the world depicted within the panels. *Little Nemo* repeatedly invests urban space with the idea of immanent collapse and a sense of falling in a way that refers to tangible vertical events, such as the construction of the skyscraper. In both series, verticality is also a manifestation of abstract ideas about the intangible realms of the human mind. But the dream-comics are not simply a refraction of the way that the human mind was conceptualized by early modernist discourse. Rather, McCay's work draws these two strands together to express a convergence of the concrete and abstract, reflecting upon both tangible changes to the world, such as the new skyscraper, and also an overturning of previous epistemologies of consciousness.

VERTICAL WORLDS: FREUD'S WORK AND *LITTLE NEMO*'S STRUCTURES OF TIME AND SPACE

Others have argued that because McCay's graphics preceded the widespread influence of Sigmund Freud's work they cannot be regarded as a refraction of Freudian psychoanalytic models.[11] Yet the spread of psychoanalysis itself was symptomatic of a wider emphasis on the human mind that came along with the cultural movement christened modernism. The similar concerns of McCay's comics, Freud's work, modernist writing, art, and cinema reveal these discourses as interdependent. McCay's work reflects a modernist preoccupation with Woolf's "dark places of psychology." The dream-comics compound the depiction of what William James called "the hidden self," a formulation that, despite finding its theoretical grounding in Freud's later work, is symptomatic of the ways in which the cultural output of the early twentieth century began to question previously accepted tenets of consciousness.[12]

---

11. Shannon, "Something Black in the American Psyche," 195–97. Unlike Shannon's, my methodology accords with the transference of analysis from author, character, or audience to the text itself. Brooks's *Psychoanalysis and Storytelling* exemplifies a similar methodology.

12. Woolf, "Modern Fiction," 192; James, "Hidden Self," 361–74.

Regardless of whether McCay had knowledge of Freud's work, the general fascination with dreaming of which the dream-comics are a manifestation was revolutionized by the publication of Freud's *The Interpretation of Dreams* in 1900. In this text, Freud proposed that the singular nature of dreams is due to a collision between our unconscious desires and the residue of events that actually occur during our waking life, making a universally applicable theory (such as a book of dreams in which specific symbols account for precise meanings regardless of the dreamer) entirely redundant. Although *The Interpretation of Dreams* was not published in the United States until 1913, by the time McCay was composing *Little Nemo,* it was becoming influential in America.[13] A. A. Brill accounts for the increasing popularity of Freud's work in the United States as the result of the fact that "nothing is as convincing as finding something in one's self which is common to all." Brill only became "a Freudian by conviction" through the practice of dream-work, and after finding "the Freudian mechanisms" in his own dreams. Further, he observed the same in others who "accepted and adhered to it only after they had found the Freudian mechanisms in themselves."[14] Brill's account suggests that American audiences embraced psychoanalysis largely because of its depiction of dreaming as an unprecedented and unrepeatable individual experience, and its corollary argument for self-narration defined by that same specificity.

Proposing self-reflection and the creation of an auto-narrative through psychoanalysis as the only way to decipher the dream content, *The Interpretation of Dreams* anticipated the evolution of later post-Freudian therapies in America that Christopher Lasch argues emphasize "the immediate gratification of every impulse." In the "age of narcissism" Lasch describes, Freudian principles are converted into a new form of American individualism in which the world is "a mirror" held up to the subject contrary to the founding principles of psychoanalysis.[15] While the emphasis on the individual in the age of narcissism can be traced back to Freud's

---

13. For a synopsis of the philosophy of dreams before Freud and its correlation to his theories, see Ferguson, *Lure of Dreams.* Freud's work was reported by the American press in articles such as "Mystery of Dreams Revealed," *San Francisco Call,* 18 March 1900, 8; "Psychology of Dreams," *Los Angeles Herald,* 24 July 1910, 16. See also Hale, Jr., *Freud and the Americans;* Rosenzweig, *Freud, Jung and Hall.*

14. Brill, "Introduction and Development of Freud's Work," 319. For details of the impact of Freud's work in America as well as Freud's visit to Clark University, Massachusetts in 1909, see Rosenzweig, *Freud, Jung and Hall,* 13; 241–43; Gay, *Freud,* 206ff.; 563; Freud, *Autobiographical Study,* 51–52.

15. Lasch, *Culture of Narcissism,* 13; 56ff.; 10. Lasch distinguishes this individualism from that tied to the idea of the American Adam (9–13).

dream-work, the idea of a privileged individual narrative is equally essential to the determining principles of *Little Nemo,* which center on, and exclusively represent, Nemo's perspective. *Little Nemo* is crucial to an understanding of not only the various discourses attached to verticality across the twentieth century but also the way that the formal and thematic energies of those narratives refract the moment out of which they are born. The cultural climate at the time of *Little Nemo*'s publication indicates that the series intersected with a new model of self-perception that would have lasting resonance; working backward from Lasch's narcissistic age, the simultaneity and complementarity of *Little Nemo*'s Slumberland (as a privately realized world unique to its protagonist) and psychoanalysis (as a new and increasingly popular mode of self-examination) indicates that both were precipitated by an emergent individualism that esteemed sui generis narrative above all else.

The tenet underlying my claim here—namely, that the artistic and scientific discourses of the fin de siècle share what Judith Ryan calls a "complex intertextuality"—is a commonplace of scholarship that regards modernism as a product of interpenetrative and mutually enriching disciplines.[16] The entire enterprise of psychoanalysis, and Freud's work in particular, engages in this intertextuality. John Brenkman describes psychoanalysis as a vacillation between autobiography and analysis in "an utterly new form of expression" with the end goal of an "autobiographical project"; psychoanalysis is an "amalgam of free association, dream, and transference" constantly "reworked by constructions, rememberings, and interpretations." In this way, Freudian psychoanalysis both worked toward and in its very methods entailed "individual self-narration."[17] Freud explicitly forges his own equation between psychoanalysis and storytelling by proposing that the unconscious mind is properly understood by deciphering symptoms as "allusions," and texts such as *Beyond the Pleasure Principle* can be, and have been, regarded as literary texts in their own right.[18] Freud's writing—a collage of autobiography, scientific theses, anecdotes, and graphics—also contributes to the explosion of experimental narrative modes that brought together disparate media in the search for a more powerful collage form, a process catalyzed, like a great deal of later modernist texts, by a preoccupation with the buried places of the psyche.

---

16. Ryan, *Vanishing Subject,* 224. See also Felman, "To Open the Question"; Micale, *Mind of Modernism;* Marcus, *Tenth Muse.*

17. Brenkman, "Freud the Modernist," 173.

18. Freud, "Manifest Content and Latent Thoughts," 100. Brooks analyzes *Beyond the Pleasure Principle* as a literary text in his *Reading for the Plot.*

Freud's work specifically echoes the formal practices of other modernist texts by drawing together verbal and visual descriptive modes. In the 1920s, Freud described "the sexual life of adult women" as "a 'dark continent' for psychology," textually echoing James's "hidden self" and Woolf's "dark places of psychology."[19] Freud also rendered verticality in visual form via an illustration that describes the id, ego, and superego (refining earlier concepts of the conscious, preconscious, and unconscious).[20] Although he would redefine the most appropriate model for representing the psyche, Freud consistently relied upon the idea of a topographical map to describe psychic processes, and the drawing of the ego subtended by the demands of the id and superego dramatizes his tendency to imagine the human mind in vertical terms. In the prose accompanying this illustration, Freud approaches psychic structure through analogies of terrain: as a "landscape of varying configuration," "hill-country, plains, and chains of lakes."[21] The id, the deepest part of the unconscious, is the "dark, inaccessible part of our personality"; what "little we know of it" is learned from the dream-work and "the construction of neurotic symptoms."[22] After the depiction of the id as a subterranean landscape, Freud then proposes another vertical analogy, invoking the spatial image for the id that implies its depths: the id is "a chaos, a cauldron full of seething excitations" in which "instinctual cathexes" (psychic energies determined by unconscious desires) swarm, seeking "discharge."[23] In his prose, Freud describes the psyche as a combined effect of how the conscious and unconscious operate under the dictates of the energy that is cathected (invested) in, and infused between, them. His written topography of the mind describes symptoms as the rerouted expression of unconscious wishes that find release in an alternate but not necessarily effective outlet. As such, Freud's vertical model makes visible an abstract concept in which unconscious wish impulses lie below the contours of a known mind.

McCay's point of interest (the individual and singular dream-narrative of Nemo) and that of psychoanalysis (the revelations that come with

---

19. Freud, "Question of Lay Analysis," 212; James, "Hidden Self," 361–74; Woolf, "Modern Fiction," 192.
20. Freud, "Dissection of the Psychical Personality," 78.
21. Ibid., 72.
22. Ibid., 73.
23. Ibid., 73–74. *Cathexis* is a term Freud coins from the German *besetzung* ("investment") to suggest the energy circulating in psychic processes; he never explicitly explains the term but comes close in *The Unconscious* (1915) and "The Neuro-Psychoses of Defence" (1894).

deciphering dreams) create a textual clash within *Little Nemo* much like Mikhail Bakhtin's heteroglossia, or the stratification of different "voices" within a text. Bakhtin suggests that the modernist novel is the form best suited to heteroglossia as it productively expresses the conflict between voices within its narrative.[24] But heteroglossia is as significant in *Little Nemo* as in the novels referred to by Bakhtin; here McCay's conscious narration of dream experience, and the model of dreaming as an unconscious process, co-inhabit the panels of the dream-comics. The convergence of different voices written into *Little Nemo*'s structure is also joined by the collision between Nemo's tangible world and the psychic one suggested by his dream experiences. Cumulatively, the coexistence of McCay's surface narration and a contemporaneous understanding of dreaming as the entrance into a hidden world make *Little Nemo* into a unique palimpsest: more than merely anticipating the modernist fascination with forms described by Laura Marcus as those of "superimposition, simultaneity, collision, dialectic,"[25] *Little Nemo*'s layers both reveal Nemo's overt narrative and reflect back upon the cultural climate of the period in which the series was produced.

The theories of psychic structure manifest in Freud's drawing of the mind and *The Ego and the Id* (1923) were precipitated by his earlier work regarding dreams and, specifically, the hypothesis that dreams are the route to the unconscious. In the Freudian model, dreaming lies at the very heart of the distortion of wish impulses in psychic processes; the dream element "is not in itself a primary and essential thing, a 'thought proper,' but a substitute for something else unknown to the person concerned." It is the frustration or incompleteness of psychic "discharge" that results in symptoms of neurosis or sublimated behavior; symptoms, "derived from the repressed," are the repressed's "representatives before the ego"; in the same way, dreams are symptoms of the id.[26] Specifically, dreams are symptoms, "the distorted surrogate for the unconscious dream thoughts," created by "the defensive forces of the ego, of resistance." The idea at the heart of a dream stands "in relation to the repressed element like an allusion, like a statement of the same thing in *indirect terms*."[27] The "distortion" in dreams that "hinders our understanding of them" is

---

24. Bakhtin, *Dialogic Imagination*.
25. Marcus, "Introduction: The Contribution of H. D.," 102.
26. Freud, "Manifest Content and Latent Thoughts," 94; "Dissection of the Psychical Personality," 74; 57.
27. Freud's "Third Lecture," reproduced in Rosenzweig, *Freud, Jung and Hall*, 420; 416 (emphasis in original); 409.

due to "the activities of a censorship" directed against the "unacceptable," and unconscious, "wish-impulses."[28] In 1890 William James had suggested myriad "pathological states" might be explained by "the existence of some perverse buried fragment of consciousness obstinately nourishing its narrow memory or delusion, and thereby inhibiting the normal flow of life."[29] Freud's model of the unconscious translates James's idea of a mentally "buried fragment" into a vertically realized representation that is both visual and textual.

If Freud's descriptions render invisible mental processes in vertical spatial terms, graphic narrative as a genre has a particular affinity for a symbolism that is reliant on the shape of vertical structures. The relationship between architecture and narrative art is embedded in the shared etymological roots of the words *story* as narrative and *story* as the floor of a building. *Story* comes from the Latin *historia,* meaning both picture and the horizontal division of a building; the meaning of *story* as a row of pictures follows from the medieval practice of positioning a picture in every window of a building.[30] In an extension of this parity, iterative images of tall buildings and other structures are frequently symbols that reflect upon, or condense, moments in a graphic narrative's storyline. *Little Nemo* rests upon vertical symbolism in the same way that Freud's model of the mind relies upon vertical analogies of an id lying below the ego, and McCay's imagery compounds the symbolic verticality of James's statement while also anticipating the visual graphic of Freud's model.

A primary tenet of Freud's work is that dreams predominantly consist of "visual images" and, beyond this, psychoanalysis creates a mosaic of ideas via the decoding of dream allusions.[31] Thomas A. Bredehoft's theory of "comics architecture," which denotes a relationship between graphic narrative's construction and the meaning imparted by its storyline, can be extended here to describe how *Little Nemo* approximates Freud's model of the dream-work. *Little Nemo*'s formal architecture makes visual the processes of condensation and displacement that in Freudian terms transform a dream's latent content (the thoughts at the root of the dream)

---

28. Freud, "Symbolism in Dreams," 125. See also Freud, "Manifest Content and Latent Dreams."
29. James, "Hidden Self," 372.
30. Raeburn, *Chris Ware,* 25–26.
31. Freud, "Difficulties and Preliminary Approach to the Subject," 78; "Manifest Content and Latent Thoughts," 100.

into its manifest content (visual images).³² As such, *Little Nemo* both emphasizes the symbolism of specific vertical structures and betrays the symbolic weight of verticality as a mode of communicating the invisible processes of the mind during dreaming. As a carnivalesque world affiliated with the realm of the unconscious, Slumberland also places a premium on the underlying aim of psychoanalytic work to document places "hidden" from consciousness; McCay's images of vertical collapse and Nemo's various ascents over skyscrapers are tied to the abstract and conceptual structures of knowledge relating to how the mind works that are manifest in both James's and Freud's conceptions of a vertically layered consciousness.

THE DREAM-COMICS, ANXIETY, AND VERTICAL DISSOLUTION

In the dream-comics, vertical imagery is most frequently tied to the expression of various disruptions and dissonances. The form and imagery of McCay's work evokes the process of converting latent to manifest content in the dream-work, while also entertaining the same attention to a vertical model of the mind and the processes existing within it. The various conflicts articulated by *Little Nemo*'s form and imagery parallel the processes of Freud's reality principle, the general thesis of which holds that the mind recognizes the inefficacy of an unrestrained urge for constant pleasure (the pleasure principle) and instead modifies this urge by accepting temporary displeasure in pursuit of long-term pleasure. This same process also ensures that what is repressed by the censorship of the ego is displaced and erupts in displeasurable symptoms.³³ The exchange between exhilaration and anxiety attached to vertical events throughout *Little Nemo* suggests that the kind of vertical activity Nemo repetitively engages in serves a double purpose as that which the protagonist both longs for and fears. As such, ascent and free fall are repeatedly deployed as if they are substitutes for other wishes that cannot be adequately realized.

Little Nemo's depiction of an overlapping circuit in which verticality mediates both discomforting and pleasurable experience also echoes Freud's description of flying and falling in dreams. Although dream

---

32. Bredehoft, "Comics Architecture, Multidimensionality, and Time," especially 869–70. Freud, "Dream-Work," 144–47.

33. Freud, *Beyond the Pleasure Principle*, 1–4; 7.

interpretation rests upon the exclusively individual meaning of dreams—each dream has a meaning contingent on the dreamer's psyche—Freud concedes that there are some "typical" dreams that seem universal and may have common roots.[34] He suggests that when the dreamer "finds himself flying through the air to the accompaniment of agreeable feelings" or "falling with anxiety," the subject is replicating "impressions of childhood," such as flying in the arms of an uncle. The condensation of childhood "romping" into flying and falling in dreams is, as in real life, accompanied by anxiety: "as every mother knows, romping among children actually ends in squabbling and tears." In a *Little Nemo* discussed later, a long and perilous slide down a banister in discrete panels breaks up time into staccato moments, generating a similar sense of the precariousness of childhood experiences that are effected by enjoyable and also unsettling moments. The pleasure of flying is not only accompanied, but in some way ensured, by the anxiety of falling. Of the typical dreams, flying, falling, and drowning are, tellingly, those which elude any closed interpretation from Freud, who professes that he is "unable to produce any complete explanation of this class of typical dreams."[35]

Given his repeated insistence on the individuality of dreams (and therein their meaning), this acknowledgement is more in keeping, in fact, with *The Interpretation of Dreams* as a whole than the generalizations Freud makes in his chapter on "typical" dreams. It also indicates the possibility of vertical dreams to be both a universal experience and one denied any final signification. The intersection between Freud's analysis of flying and falling in dreams and his model of repressed impulses seeking discharge suggests that verticality is a reflection of some other anxiety or desire. Both the depiction of falling and flying as jarring experiences in *Little Nemo*'s narrative content, and the series' formal practices, are essential components of the way in which McCay's work anticipates the overtly vertical imagery of psychoanalysis's depiction of the delayed eruption of what is repressed. Furthermore, *Little Nemo* strips such as the very first one, in which the panels' progressively disruptive imagery and darkening colors create a sense of visual claustrophobia, transmit an energy not unlike the increasing pressure of the forces described by Freud's description of repressed impulses that have been buried only to incessantly seek a release.

---

34. Freud, *Interpretation of Dreams*, 339–80. See also Bachelard, *Air and Dreams*, 19ff.

35. Freud, *Interpretation of Dreams*, 374–77.

Little Nemo can be read most productively as the transmission of broader modernist interrogations about individual consciousness, as well as a text that reflects what can be called a cultural unconscious. Specifically, the dream-comics flirt with certain cultural taboos in a way that correlates with the construction of a "group mind."[36] Essential to this is the very repetitiveness of narrative plot in, and the sequential daily publication of, the dream-comics. Obeying the same falling asleep/dreaming/waking tripartite structure in each strip, the dream-comics also provided a daily act of consumption for their readers, inscribing a repetitiveness both in the series' internal plot and their reception by an audience. The continual and cyclical repetitions of *Little Nemo* create a textual energy that has much in common with the model of repression; the suppressed element in *Little Nemo* reemerges in a different form in a manner comparable to the dictates of the censorship of the ego. Repetition is essential both to the continual eruption of the wishes of the unconscious into the conscious and the "compulsion to repeat," a puzzle that prompts Freud to consider whether repetition is the way in which the ego tries to avoid the displeasure that would occur with the "liberation" of what is repressed, but in so doing, gets caught in a loop.[37]

McCay's work formally and conceptually mobilizes dreaming as a state in which spectacular, and frequently threatening, vertical events communicate a sense of instability. Slumberland both indulges that sense of an "impossible," alternate world of dreaming and makes that world constantly bump against a waking reality, transforming it into a metaphor for the fragmentation of subjective reality. Like Freud's suggestion that dreams are the route to the unconscious, the action within *Little Nemo* is a kind of oscillation between waking and dreaming states that interact with, and inform, one another. As such, Nemo's dream-world evokes an experience analogous to schizophrenia, drug-induced hallucinations, or paranoiac fantasy, suggesting that Slumberland is a substitute for the desire to experience an alternate psychic state, a condensation of the issues of mental health and race suppressed by the mores of society.

It is the articulation of alternative consciousness and the entrance of the racial other in McCay's dream-comics that augur the most extraordinary images of vertical disorder and collapse. *Rarebit Fiend* in particular

---

36. See Freud, *Group Psychology*. Here, Freud exposes the ways in which groups form a collective mind that exhibits different "ideas and feelings" than those of the individual mind, and how characteristics of "contagion" ensure that the individual often surrenders to collective urges (7; 10ff.).

37. Freud, *Beyond the Pleasure Principle*, 14.

flirts with a grotesque world in which verticality both represents the unlikely and physically dislocating experience of dreaming and also creates the foundation for metaphors relating to mental disturbance. Although *Rarebit Fiend* does not have the sustained narrative continuity of *Little Nemo*—rather than the protagonist Nemo, to whom all the experiences occur, each strip presents a new character—the earlier series indicates McCay's burgeoning fascination with dreaming as a narrative tool that permits fantastic and disturbing storylines within the legitimate frame of a waking narrative. A strip published on 25 February 1905 assumes the perspective of a corpse being buried; the dreamer awakes in the final panel believing that clods of earth are being thrown down upon him as he lies at the bottom of the grave. As in *Little Nemo,* the strip's final panel seems to distinguish between reality and dreaming while simultaneously implying a disquieting blur between the two states. Here the dreamer's exclamatory cries fracture his apparent relief; he wakes to realize he is not actually lying dead in a grave, but his calls, "Oh! Shucks! It's a dream, good!" and "Oh! Oh! What a dream! Oh," are so repetitively punctuated by the inarticulate "Oh!" that they suggest a continued feeling of distress in spite of this realization.

By reconstructing the viewpoint of the dreamer, the perspective of each panel not merely invites the reader to assume the position of being underground and therefore share the claustrophobia of the protagonist but also textually attaches a concrete orientation of up or down to a vertical metaphor. In the fourth panel a sign attached to the dreamer's grave reads, "HE LIVED TOO HIGH," while one of the mourners states that "he looks natural but who wouldn't with the alcohol in him." The doubled meaning of "high" as both an elevated position and, colloquially, to be intoxicated makes the fourth panel read as an allusion to alcoholism. It also creates a visual joke predicated on the fact that in his imagined death the protagonist is literally below ground, as if in perfect inversion of his metaphorical "highness" while alive and drunk. As such, the strip visually relies upon the motion of going down and the perspective of looking up, and textually turns upon a vertical metaphor, combining its written and illustrative energy to articulate a perturbing dream made no less so by the final return to waking life. The experience of being lowered into a grave also facilitates a crossover between dreaming and the experience of cognitive difference. As throughout the dream-comics, if the episode recounted was experienced while awake, it would be considered a hallucination or an indication of mental disturbance; dreaming is both different from, and a permissible relative of, psychological illness.

Rarebit Fiend communicates the subjugation of narratives concerning alternative states of mind only for these same narratives to reemerge under a different guise, crucially, via the disquieting tenor of vertical imagery. Furthermore, the representation of alternative consciousness in the dream-comics suggests that death and sleep are not entirely independent states, compounded by *Little Nemo*'s invocation of Morpheus who, in the classical relegation of the myth, is related to the twin gods of sleep and death, Hypnos and Thanatos. The suggestion that dreaming is both a gateway to other states of consciousness and a close relative to death furnishes the final panel of the dream-comics with an enhanced sense of uncertainty. Here the final crossover into real life is haunted by not only the leakage of dreaming into reality but also the concomitant implication that, with that leakage, comes a crossover of mental disturbance or even death into a "safe" waking world. In *Rarebit Fiend,* the anxieties about cognitive difference communicated via the startling incident of being lowered into the ground are joined by moments in which the force of racial otherness generates comparable moments of disruption.

In fact, *Rarebit Fiend* and *Little Nemo* reflect an overt racism that, at the time, was comparatively permissible in the popular press and distracts from less socially acceptable anxieties so central to McCay's work, such as those surrounding mental illness.[38] Both series frequently use humor to facilitate racist anxiety, suggesting that humor as a form of what in psychoanalytic terms is called a *displacement* or *condensation* of repressed impulses.[39] In a strip published on 17 January 1905, an African American maid dreams that she can alter her skin tone with bleach and after doing so is mistaken for the white "Duchess of Oshkosh," only to awaken decrying the return to reality as "a scandalous shame." The strip's guileless humor seems a displacement of very real racial hostility, and the black figure is no more elaborate than a golliwog doll or blackface minstrel, both of which emerged at the cusp of increasing malevolence directed at African Americans. The golliwog doll has a particularly insidious origin; it was a toy of increasing popularity in the early

---

38. For a full discussion of race and early comics with regards to McCay, see Roeder, *Wide Awake in Slumberland,* 69–74; 88–90. Shannon also discusses the representation of race in McCay's work, concluding that McCay's work is "essential" to understanding American comics' "repressed narratives of race, power, and privilege" ("Something Black in the American Psyche," 211). Shannon's claims are similar to mine, but he prefers to emphasize McCay as "non-Freudian" rather than interrogate the relationship between McCay's work and the modes of thought that came about because of Freud's work.

39. Freud, *Jokes.*

twentieth century, inspired by a character described as "a horrid sight! The blackest gnome" in Florence Kate Upton's 1895 children's book, *The Adventures of Two Dutch Dolls and a Golliwogg*.[40] The stereotyped figure also ushers in a structural destabilization as the usual format of the dream-comics—in which the final return to waking life is unsettling yet seems a temporary relief from the frightening experiences held by the dreamworld—is reversed: here the dreamer longs for the possibilities of sleep rather than the reality of waking up. In order to facilitate the perspective of an African American, it is as if the strip must overturn its usual procedures, suggesting that an even larger mental shift must take place in order for the narrative of another race to be considered.

William J. T. Mitchell argues that pictorial representations occurring at the interstices between groups of people—that is, the "stereotypes, the caricatures, the peremptory, prejudicial images that mediate"—can "take on a life on their own," a "deadly, dangerous life," which exists in the "rituals of the racist (or sexist) encounter." Moreover, "precisely because the status of these images is so slippery and mobile, ranging from phenomenological universes, cognitive templates for categories of otherness, to virulently prejudicial distortions," their "life" is "difficult to contain." The "life" Mitchell describes ensures the seemingly unstoppable proliferation of the stereotype that can make interracial face-to-face encounters fraught. The repetition of pictorial stereotypes in McCay's comic strips, like the golliwog doll and the minstrel in real life, makes an encounter between races within the narrative "never unmediated," makes it charged "with the anxiety of misrecognition and riddled with narcissistic and aggressive fantasy."[41] The ridicule attached to the minstrel-like caricature contributes to a wider cultural process in which the reinscription of racism produces more anxiety, so that the underlying fear never becomes resolved.

Both the specific stereotypes in McCay's work and wider stereotypes, such as the golliwog doll, approximate the Freudian model of an unconscious repressed satisfaction that manifests itself in an inadequate discharge—a kind of eruption from the "deep" part of the psyche. Furthermore, the existence of racist stereotypes in *Little Nemo* combines with moments of vertical disorientation that culminate in the complete disintegration of a vertical world, as if the inefficacy of what is being repressed reaches such a force that the very structure of Slumberland itself

---

40. Upton, *Two Dutch Dolls*, 6.
41. Mitchell, *What Do Pictures Want?*, 297; 295–96.

must collapse. Crucially, these moments in *Little Nemo* are ushered in by the entrance of the "primitive," a figure evolved from McCay's cartoons in *Life* magazine, contributing to the popular press's simultaneously antiimperialist and racially ambivalent attitude. Exemplary is a political cartoon in *Life* from 23 November 1899 by McCay in which Uncle Sam is tied to a tree entitled "Imperialism" by a bucking donkey entitled "Philippines," while an old man inscribed as "Spain" wanders over the horizon out of reach. While the image casts the imperial enterprise as a foreboding and difficult one, it nonetheless depicts the Filipino people in derogative, animalistic terms. American political cartoons during the Spanish-American War in general indicate the movement from a previous era of paternalistic racism into one bolstered by ambivalence that would soon become manifest in the increasingly pernicious attitude toward African Americans in the early twentieth century. Declared in 1898, the Spanish-American War was ostensibly a response to Spanish atrocities in Cuba, but its result—America's colonial possession of Puerto Rico, Guam, and the Philippines—uncovers how it was nurtured by military adventurism. The myth that William Hearst in fact facilitated the Spanish-American War, and even promised the artist Frederic Remington that he would "furnish the war," has largely been discredited by recent scholarship, but yellow journalism undoubtedly played its part in a particularly new meditation on American imperial encounters.[42]

Significantly, the earlier ambivalence of McCay's political cartoons transforms in *The Jungle Imps, Rarebit Fiend,* and *Little Nemo,* all of which feature overt stereotypes of African Americans, Native Americans, Chinese, and unidentified "primitives" in hostile guises. A collaborative enterprise between McCay and George Randolph Chester printed in 1903, *The Jungle Imps* was a spoof of Rudyard Kipling's *Just So Stories,* which had been published the year before.[43] Kipling had famously exhorted Western imperial designs and simultaneously posed them as hugely problematic in the 1899 poem, "The White Man's Burden: The United States and the Philippines Islands," but *Imps* flippantly recasts Kipling's work, dismissing the perspective of non-American nations and paying only cursory attention to the problems of imperialist venture. It features a horde of black figures, including the recurring characters Gack, Boo-Boo, and Hickey, who, despite their innocuous names, harbor brutal intentions. In each episode, the imps torture an animal, only for

---

42. See, for example, Campbell, *Yellow Journalism*.
43. Robb, "Winsor McCay," 247–48.

the same animal to undergo surgery at the hands of "Doctor Monk," who recreates its form and abilities, and allows it to exact revenge upon the imps. The recurring narrative structure, as in *Little Nemo*, cements *Imps* as a constant reinstatement of the same ideas: most especially, the consistent thematic denigration of cultures imagined as "primitive."

The same caricatures of *The Jungle Imps* reemerge in *Little Nemo*, implying that Nemo's Slumberland incorporates the world of the imps, which is not presented as a dreamworld, per se, but rather a fantastic version of the real world in which animals are given sentience. The leakage of characters between the two strips further destabilizes that distinction between what is dreaming and what is waking life, as well as suggesting that *Little Nemo* contains the same anxieties surrounding race that are so central to *Imps*.[44] One imp, simply entitled "Impy," becomes a recurring companion to Nemo in his dreams after a visit to the "Candy Islands." In the strip published on 30 August 1908, Nemo finds himself with Impy in yet another jungle world, where they encounter yet another tribal culture. The ominous characters in *Imps* expand into *Little Nemo*'s depiction of an even more frightening people; the imps' cruelty directed at animals in the former comic strip transmutes into the cannibalistic intentions of the latter's tribal people. The tribal figures are extensions of the jungle imps in more ways than one: both an elaboration of the same racist ideas they symbolize and a literal enhancement of their size. The fear associated with the racial other, temporarily held at bay by *Imps*, explodes in *Little Nemo* as a disproportionate fantasy of primitive ritual centered upon the capture of a vulnerable white boy that is symbolically compounded in the vertical growth of the imps.

Furthering the vertical distortion at work in this strip, the one published on 26 June 1910 demonstrates the force of racial difference both as it is represented within the strip's panels and in the effect it generates upon the structure of the sequence. Panels 5–8 of the original drawing for this strip present the culmination of a story involving Nemo, Impy, and Flip, a figure who was possibly the representation of a "rotund black man" McCay saw smoking a cigar in Brooklyn.[45] Each figure falls in front of a backdrop of vertical plant stems.[46] After a bee attacks Impy,

---

44. For other readings of the anxiety of race in these strips, see Honeyman, "Gastronomic Utopias," 58–59; and Shannon, review of *Little Nemo in Slumberland*, 97–98.

45. Canemaker, *Winsor McCay*, 113.

46. The original drawing of this strip is held by the Library of Congress Prints and Photographs Division and is viewable both on-site and online, under the title "Little

Flip kills it with his cigar. The sequence establishes a shape of frame in which a vertical panorama enforces an up-down reading practice within the limits of each panel. The emphasis on a vertical panel and a vertical process of reading is abetted by the vertical motion (Nemo's fall) and the vertical structures (the plants which form the background) depicted in each of those panels. What Flip and Impy signify within the parameters of *Little Nemo* as a whole lends particular significance to specific instances such as these, in which verticality governs not merely the action depicted but the mode of reading that action. In McCay's work, an underlying force attached to the racial figure appears to engender disturbing vertical events, such as a drawn-out fall from space. These images of vertical collapse and disorientation in the dream-comics can be read as visual manifestations of what the narrative communicates about the disquiet prompted by cognitive or racial difference.

PAPER BUILDINGS: COMICS ARCHITECTURE AND THE COLLAPSING CITY

McCay's turn-of-the-century aesthetic is preoccupied by what I referred to in the introduction of this book as the *vertical frontier*. *Little Nemo*'s heavy reliance on vertical geographies refracts the growth upward of the American city in the same years as its publication, after the closure of the frontier and the ostensible end of horizontal space in a westward direction that it signaled. In its reflection of this new and metaphorical vertical frontier, *Little Nemo*'s modernist aesthetic also fuses low and high cultural indices, blending the popular comic strip form with often esoteric architectural images. Scott Bukatman suggests that *Little Nemo*'s more obscure architectural allusions summon "the ornate and singular façades on the apartment buildings of the *belle époque*."[47] But this observation does not fully account for Slumberland's architectural shapes, which variously recall both the urban skyscraper and also the shapes of the 1893 Chicago World's Fair, as well as the vertical structures of the popular Coney Island amusement parks, Luna Park and Dreamland (opened in 1903 and 1904, respectively).[48] These same architectural shapes are often subject to vertical collapse in tableaux that further attach Slumberland

Nemo in Slumberland. Oug! oug! glabble gick iggle gouck" (26 June 1910, drawing of ink over pencil with paste-on, holding number LC-DIG-ppmsca-13287).

    47. Bukatman, *Matters of Gravity*, 187. See also Canemaker, *Winsor McCay*, 100.
    48. Roeder, *Wide Awake in Slumberland*, 108–12.

to the amusement park, recalling Coney Island's staged destruction of buildings in "disaster spectacles."⁴⁹ In addition to the visual vocabulary of Slumberland's buildings, Dreamland has several further influences on *Little Nemo*; the compound noun *Slumberland* verbally echoes *Dreamland,* and the depiction of Nemo sliding down banisters or flying through the air, as well as his encounters with strange characters and stereotyped African American figures, could all legitimately come from Dreamland's rides, "freak show" exhibits, and vaudeville theater. While Slumberland's visual coordinates draw together the real world with a fantastic one, the vertical events within *Little Nemo* become a vehicle for the crossover between banal city life and the extraordinary world of dreaming. *Little Nemo*'s allusions to amusement parks offer what Katherine Roeder calls "the dizzying perspectives" of new experiences such as urban transport and roller coaster rides.⁵⁰ By synthesizing the spatiality of an everyday and legitimate urban experience with the fantastic and unpoliced realm of the amusement park, *Little Nemo* encourages a conversation between the demands of logic and illogic that are placed upon Nemo's experiences.

In a broader sense, *Little Nemo*'s representation of urban vertical structures has an important place in the emergence of a modernist conception of the city. In the 1910s and 1920s, the new urban landscape was represented, interrogated, and, to different extents, championed by American literature, painting, photography, cinema, and cultural criticism, and was regarded as a sign of ruthless capitalism or one of possibility and freedom. The skyscraper emerged in Chicago after the fire of 1871, after which architects were faced with what Robert Hughes calls a "blank slate." By the time the frontier was closed, the first American skyscrapers were already in the process of claiming that "unowned and previously useless space, the sky itself."⁵¹ There is, however, a comparable absence of the skyscraper in texts before the turn of the century; it is as though the frontier held such a monopoly over spatial metaphor that it was not until the declarative "end" of the frontier that there emerged representations of the vertical that might be expected with the birth of the tall building. The prominence of the skyscraper in narratives only after 1900 is perhaps best

---

49. See Dennett and Warnke, "Disaster Spectacles." I discuss the disaster spectacle and amusement park in greater detail in chapter 2.
50. Roeder, *Wide Awake in Slumberland,* 102.
51. Hughes, *Shock of the New,* 172; 10. See also van Leeuwen, *Skyward Trend of Thought*; Mumford, *City in History* and *Brown Decades;* Landau and Condit, *Rise of the New York Skyscraper*; Moudry, *American Skyscraper*; Schleier, *Skyscraper in American Art*; Nye, *American Technological Sublime,* 88–89.

understood not exclusively in terms of the end of the frontier, however, but also within the wider context of other shifting contemporary conceptions of space. In this sense, the way in which the skyscraper comes to haunt literary and visual texts is colored by the impact of various spatial revolutions. These revolutions form a congregation Lewis Mumford glosses as comprising "new devices of spatial liberation" (the automobile, the airplane, and the radio); revelations that the atom held "unsuspected complexities"; and the idea that "hitherto untouched depths in the mind" existed.[52]

In a way, New York condenses the manner in which the entire nation found itself bound by the limits of its Pacific coast at the end of the nineteenth century, and the city's unique claustrophobia—its circumscription via a shoreline that places a premium on vertical space—is essential to early depictions of the newly vertical city. Although Chicago was the site of the first American skyscrapers, New York is the most pervasive and popular site of the city narrative in the first quarter of the twentieth century. The city's prominence in narrative is perhaps the product of the particularly disjunctive spatial aesthetic created by its first skyscrapers, an aesthetic that was influenced by the dynamics of economic and cultural change at the fin de siècle, and which broke away from Manhattan's theoretically rational and ordered two-dimensional grid system into a skyline of irregular skyscrapers erupting into three-dimensional space.[53] New York is also historically the most popular site of fictional large-scale disaster or apocalypse, a pattern Max Page considers in terms of the cycles of destruction and rebuilding that take place in Manhattan, and which bears special consideration in regards to the images of collapsing buildings in McCay's work.[54]

In 1896 Louis H. Sullivan proposed that the skyscraper must contain "the force and power of altitude," and that it "must be every inch a proud and soaring thing, rising in sheer exultation."[55] Lewis Mumford would later dismantle Sullivan's article, suggesting that it makes "mischief" via "the notion that on the foundation of practical needs the skyscraper could or should be translated into a 'proud and soaring thing.'" Despite Mumford's complaint—that Sullivan fails to acknowledge the practical impetus to create tall buildings and instead gives the skyscraper

---

52. Mumford, *Brown Decades*, 64; 113.

53. Koolhaas, *Delirious New York*, especially 104. See also Scobey, *Empire City*.

54. Page, *Creative Destruction of Manhattan*; idem, *City's End*. See also Yablon, "Metropolitan Life in Ruins."

55. Sullivan, "Tall Office Building Artistically Considered," 406.

"a spiritual function to perform"—Sullivan's words represent precisely the significance of verticality in American narrative after the turn of the century.[56] Sullivan may exclude the specific economic and historical conditions that have enabled the tall building to be constructed, but, in so doing, he reflects a desire to make the aesthetic reach of the skyscraper into a meditation on what that same height might say about American thought and culture. The conflation of a physical vertical object with its various associations—whether aspirational as in Sullivan's words or anxious in post-9/11 discourse—reveals the seam between visible space and the meanings attached to it. After 9/11 the critic Philip Nobel implied that architects planning the rebuilding of the Ground Zero space were being asked to do something unprecedented: to "make buildings speak, give them meaning, create symbols from a culture with no common code."[57] However, the negotiation between material objects and the intangible ideas they conjure is not, as Nobel suggests, a new problem, but one that is both salient and revealing in American texts after the turn of the century.

Robert Hughes suggests that New York is the American city that most represents modernity, not because the city provides "a higher proportion of the canonical buildings of the modern movement than other cities," but rather because of its "Promethean verticality, its metal-and-glass severity, and its incredible power as a transformer of information and desire." From 1890 to 1930, when the vocabulary of architecture altered more substantially than it had in the previous four centuries, the paragon of "social transformation through architecture and design" became an imperative for modernism.[58] In New York, the collision between emerging modernist thought and the city's claustrophobic spatiality summons a dialogue between the new and disorienting ideas about subjective reality and dissonant physical space. This emerges in multiple and often interconnected ways in *Little Nemo* and *Rarebit Fiend*. Strips such as the *Rarebit Fiend* strip of 17 April 1908 and the *Little Nemo* edition from 29 September 1907 provide manifestations of the wider trend of representing New York as a city of ruins and simultaneously allude to the amusement park disaster spectacle discussed in the next chapter of this book.

The dream-comics are often strikingly similar to the representation of ruin in post-9/11 texts. They are also deeply embedded in the

---

56. Mumford, *Brown Decades*, 69–70.
57. Nobel, *Sixteen Acres*, 7.
58. Hughes, *Shock of the New*, 164–65.

representation of time, and the temporal and spatial structures at work in *Little Nemo* cultivate a thematic dissonance reflecting back upon early twentieth-century discussions about the implications of new ideas and technologies signaled by both new models of consciousness and the changing shape of urban space. The *Rarebit Fiend* strip published on 17 April 1908 begins with a child playing with ordinary household objects, only for these objects to transform into the vocabulary of the city—enormous vertical buildings—and then collapse in an inky rendering of destruction. The radical disintegration of skyscrapers is ostensibly an act of playful childhood imagining, but—like many other incidents in both series—it also joins the representation of changing architecture within comic frames with the structure of the strip itself. The same is true of four *Little Nemo* strips running over one month in 1907, in which the spectacular disintegration of a city skyline provides the subject of the climactic panels.[59] In the first strip of this four-strip narrative, Nemo and Impy are running from giants in the previous jungle world and enter a miniature city where they stand so tall they can clamber over skyscrapers as if in a playground. In his quest to find King Morpheus, Nemo has become "lost," and he crawls up a skyscraper to see more clearly. After his ascent, he is only greeted by more and more skyscrapers, and the *mise en abyme* of endless verticality makes him increasingly disoriented.[60] In the second strip, Nemo climbs down, fearing his pursuers, and eventually finds a river (figure 4). As he starts to escape, he sees Flip reappear, running toward him and making buildings fall down in his wake. In the third strip, the falling buildings generate a calamitous fire and by the final panel only the girders of the skyscrapers remain. The fourth strip shows Nemo and his companions fleeing across the water as the destroyed city recedes into the distance.

In this series of strips, the collapse of the series' usual structure and the vertical collapse depicted within the panels of the final strip suggest that the combined force of Impy and Flip can no longer contain the racial anxieties they represent; vertical collapse seems to be the direct result of the racial difference within their frames. The racial stereotypes in the dream-comics can be read as akin to the "Western fetishization of

---

59. McCay, *Little Nemo*, 110–13.

60. I use the term *mise en abyme* here to designate its two-dimensional meaning of the reproduction of an image into infinity, while also deliberately calling on its other meaning in literary theory, which broadly refers to the replication of themes or symbols within a text, or specifically relates to the intertextuality of language, which never reaches a core meaning due to its endless chain of signification.

FIGURE 4. *Little Nemo in Slumberland,* Winsor McCay, 29 September 1907. Courtesy of Peter Maresca, Sunday Press.

the Other" in which "natural or generic categories" are substituted for those that are "socially or ideologically determined," endowing the Other with "evil characteristics and habits" that are "inherent."[61] While this fetishism is "fixe[d] in time and place" by "commemorating a founding moment in the etiology of consciousness" and reaching back as a "memorial" (Freud's term) to an "unrepeatable first form," it is also subject to compulsive repetition.[62] In the same way, the inadequate and repetitive displacement of the repressed element seems to generate the various repetitions of the *Little Nemo* strips showing a collapsing city. As the condensation of a repressed desire to fix the racial Other in a stereotype, the collapse of *Little Nemo*'s internal cityscape, as well as the disintegration of its formal apparatus, suggests the inherent instability of that same stereotype.

In the *Rarebit Fiend* strip in which the child imagines a city falling down, the careful, regular structure of panel size is counterposed by the wild angles of what is occurring within them, so that the regularity of panel size stands in for the supposed circumscription of dreaming life: what happens in the child's dream is delimited by the panel borders. In the later series of four *Little Nemo* strips represented by figure 4, this becomes less true. These strips dissent from the usual structure that poses a new landscape in each installment, creating a short internal storyline in which Nemo falls asleep only to return to the very same landscape he has visited in the previous strip, despite seeming to have broken the dream's spell by waking. *Little Nemo*'s ruined city strips convey a return to a traumatic spectacle compounded by the basic falling asleep/dreaming/waking structure of the entire series. By echoing the dynamics of trauma, the four strips suggest the disrupted temporality inherent to Freud's *nachträglichkeit,* in which "the past is depicted from the understanding of the present instead of being kept and simply discovered in the memory."[63] This only enhances the apparent similarity between the images of destroyed skyscrapers and post-9/11 imagery, but the artistic rendition of a collapsing city recurs in various forms throughout modernism, and the convergence between post-9/11 texts and McCay's work, in fact, reveals how images of vertical dissolution are a harness for various anxieties attached to, but not always about, the skyscraper.

---

61. JanMohamed, "Economy of Manichean Allegory," 86.
62. Apter and Pietz, *Fetishism as Cultural Discourse,* 4. See also Freud, "Fetishism."
63. Eickhoff, "On *Nachträglichkeit,*" 1454. See Lacan, *Ecrits,* 33ff. See also Laplanche, "Notes on Afterwardsness"; Wyatt, "*Love*'s Time and the Reader."

Almost a hundred years after McCay's strips, Art Spiegelman's graphic narrative *In the Shadow of No Towers* forged an explicit relationship between 9/11 and McCay's earlier depictions of vertical dissolution. While *In the Shadow*'s appendix and introduction announce the relationship between the location and events of McCay's imagined collapsing city and the ruined Twin Towers in 2001, it also reframes *Little Nemo* by blending the style of early twentieth-century comics with that of contemporary work. Spiegelman's appropriation of *Little Nemo*'s narrative structure is no less relevant; one double-spread page from *In the Shadow* adopts McCay's final panel and renders it in the style of *Little Nemo*, with a mouse (alluding to Spiegelman's *Maus* in which Jews are represented as mice) on the floor while a motherly figure stands by its bed, wearing a gas mask.[64] The paranoia associated with the gas mask makes the reference to McCay's work both nostalgic and anxious, and betrays the overall tenor of a page in which bright colors jostle with one another in a way reminiscent of Hieronymus Bosch's medieval depictions of hell. Running the length of the page is the image of a disintegrating tower superimposed with a falling figure of Spiegelman himself.

Here vertical and horizontal reading practices blend so that the process of comprehending *In the Shadow* is multidirectional, proceeding from left to right as well as up to down in the long vertical panel on the left-hand side of the page, much like McCay's vertical panoramas in the one published on 26 June 1910 discussed earlier in this chapter. Throughout, *In the Shadow* manipulates dimensionality and perspective in various ways; at one point, a panel torques on a vertical axis, making a two-dimensional panel into a rendition of a three-dimensional structure that evokes the shape of the Twin Towers. The conflicts generated by visual and verbal modes of representing 9/11 in *In the Shadow* approximate the disorientation and disruption left by the experience of witnessing the Twin Towers disintegrate.[65] By adopting the structure of McCay's narrative and augmenting it with his own manipulations of three-dimensionality, Spiegelman suggests that the experience generated by the destabilization of reality and dreaming in *Little Nemo* is similar to that left by seeing the Twin Towers collapse. The depiction of a collapsing city in *Little Nemo*, transformed by the events of 11 September 2001, is bestowed with new meaning in texts such as *In the Shadow* that seek a way to work through the newly disturbing representations of collapsing skyscrapers after 9/11.

---

64. Spiegelman, *In the Shadow of No Towers*, 6.
65. Ibid., 2; Mackay, "Representing 9/11," 5–7; Chute, "Temporality and Seriality."

And yet, while in general terms, the destruction left by 9/11 appears to be presaged by earlier depictions of a ruined city, *In the Shadow* does not simply fulfill the strains of earlier apocalyptic fiction that implies collapse to be inherent to the skyscraper (discussed further in chapter 5 of this book). Where McCay's narrative structure summons a sense of disorientation or instability in its temporal disruptions and vertical imagery, Spiegelman appropriates the same modes of formal and conceptual dissolution. As such, *In the Shadow* draws to the fore the immanent collapse that is always a possibility in the very building of a skyscraper, making images of collapse not simply a return to the events of 11 September 2001, but creating the sense that verticality is a harness for more deeply held anxieties stretching back to the genesis of the first tall buildings. Furthermore, *In the Shadow* effects a sense of disrupted time at the level of the narrative and also directs its attention at an understanding of the events leading up to 9/11; it summons McCay's work in order to push against the temporal fixedness of a discourse that makes the events of 2001 into a fracture of continuity after which nothing is imagined to be "the same." Because of this strategy, when placed alongside *Little Nemo*, *In the Shadow* operates as a kind of retroactive reimagining in which the original scene depicted by McCay can draw away some of the short-sightedness that might be involved in viewing 9/11 as a watershed moment.

Little Nemo's falling figures and collapsing buildings reflect the modernist representation of the skyscraper as the simultaneous source of anxiety and awe via the promise of an always imminent collapse. The doubled pleasure and displeasure of vertical experiences in *Little Nemo* also parallel the dynamics of the sublime experience so common in modernist representations of the skyscraper. However, the frequency with which the sublime enters skyscraper narratives, and the fact that images of vertical collapse are not merely confined to post-9/11 texts, does not fully explain why verticality is central to both the form and metaphors of work from McCay to Spiegelman. While 9/11 may have prompted the recurrence of verticality as a method of expressing a particular anxiety, its origins in McCay's work have more to do with the way those anxieties articulate a modernist aesthetic. Regarding McCay's work within these parameters does not preclude an understanding of the intertextual dialogue generated by narratives that, while created nearly a hundred years apart, equally annex verticality to a pervasive sense of anxiety, but it uncovers the specifics of verticality in those separate narratives.

Precisely because they draw together the formal disruptiveness of a new kind of comics with an internal narrative intent to displace the normalcy and regularity of the everyday, McCay's comics uniquely interrogate the supposed coherence and impermeability of waking life. The collision between a material and a psychic world experienced by Nemo is mediated by representations of vertical buildings contained by individual panels, but these same frames also coalesce into the aggregate silhouette of a building, making verticality *Little Nemo*'s internal and external spatial structure. As Chris Ware suggests, the panels of graphic narrative can be treated as independent or cumulative; while one way "to aesthetically experience comics" is "beat by beat," "as you would playing music," another method is "to pull back and consider the composition all at once, as you would the façade of a building" and view a comic "as you would look at a structure that you could turn around in your mind and see all sides of at once."[66] In these terms, each strip of the *Little Nemo* series becomes a collection of separate blocks that can be read either in a way that adheres to temporal sequence (as suggested by Ware's musical analogy in which the reader experiences panels "beat by beat") or in a way that jettisons the rigidity of sequence and instead strives to comprehend the totality of the page.

As well as the two possible and discrete modes of reading discussed by Ware, *Little Nemo*'s representations are determined by other temporal and spatial practices unique to graphic narrative. While *Little Nemo* in some ways echoes the visual translation of ideas to images that is essential to Freudian dream-work, the series complicates the straightforwardness of this equation by incorporating both visual and textual information. The cross-fertilization of text and graphics—which Marianne Hirsch names "visual-verbal biocularity" and Hillary Chute calls "two narrative tracks"—governs the way that graphic narrative constructs meaning, as well as allowing it to set up a formal juxtaposition between word and image. The significance—and the uniqueness—of this duality is a central tenet of comics theory, following Scott McCloud's seminal *Understanding Comics*.[67] While there is a specific correlation between *Little Nemo* and work by Freud produced after McCay conceived his dream-comics, there is more significantly a relationship between *Little Nemo* and contemporaneous ideas that were to some extent subsumed by Freud's theories. The formal interdependence of word and graphic in *Little Nemo,* in which verbal and visual information coexist but cannot be simultaneously

---

66. Ware, quoted in Raeburn, *Chris Ware,* 25.
67. Hirsch, "Collateral Damage," 1213; Chute, "Comics as Literature?," 452. McCloud, *Understanding Comics.* Will Eisner, another prominent graphic artist, has also produced noteworthy examinations of how the form functions; see, for example, *Comics and Sequential Art.*

appreciated, specifically refracts the newly pervasive concepts of consciousness as a composition of states of mind that exist in constant flux but are yoked together in a "stream."[68] As such, the formal structure of *Little Nemo* is a crucial part of the interaction between the overarching concerns of the series and those of modernism.

John Brenkman describes modernism's impact in terms of an "imperative of newness" that demanded art "measure up as a response to the *unprecedentedness* of modern life itself, its continual transformations and dislocations."[69] *Little Nemo* responds to precisely this demand by creating a narrative that in one sense fluidly moves from one panel to another in a contiguous procession, but in another way is potentially disrupted by the division between word and image, while also being interrupted by the split of the gutter between panels. The gutter is commonly understood as the space into which the "willing and conscious collaborator" (that is, the reader) inserts their own understanding in order to comprehend the overall narrative at play. Scott McCloud calls the gutter a "fracture" of time and space that allows us to transform a "jagged ... rhythm" of "unconnected moments" into a "continuous, unified reality." Similarly, Thierry Groensteen suggests that the spaces between panels are "holes" that "project" us into a "consistent" world.[70]

This theory holds fast in strips such as that published on 18 April 1909 (figure 5), where the panels in question are distinct and offer a clear progression through time, if not through space. In this sequence, Nemo slides down a banister; each successive panel shows him in a different place, but through the process McCloud calls "closure," we infer that the banister displayed in each panel is the same one. The effect is of a consistent reality which, while ostensibly stable, does not detract from the tension generated by the actual act engaged in by Nemo; the always imminent expectation of a fall from the banister ensures a sense of precariousness. While in figure 5, a unity of space and time is upheld despite different angles of perspective in each panel, elsewhere *Little Nemo* withholds the construction of meaning through the assumption that successive panels move forward through time. Exemplary is the strip from 27 January 1907 (figure 6), in which temporal and spatial disorientation are inextricable from the verticality represented within the panels. Nemo and his companions enter the Ice Palace on their way to meet Jack Frost and, in each

---

68. James, "Stream of Thought," 224ff.
69. Brenkman, "Freud the Modernist," 173 (emphasis in original).
70. McCloud, *Understanding Comics*, 65–67; Groensteen, *System of Comics*, 10.

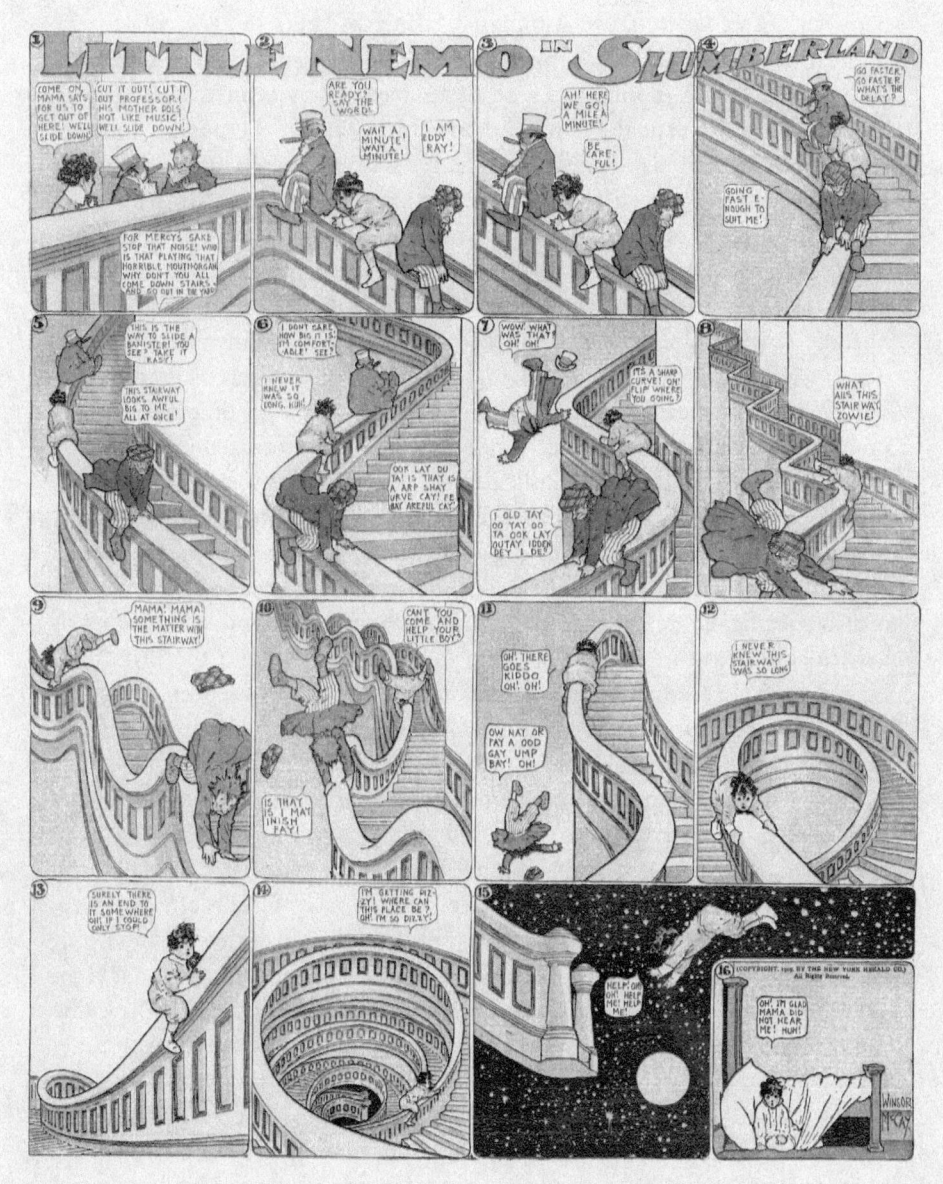

FIGURE 5. *Little Nemo in Slumberland*, Winsor McCay, 18 April 1909. Courtesy of Peter Maresca, Sunday Press.

panel, are represented a little further up a curling stairway. The contents of each frame seem to spill from one into the next with only a small space between them, so that the cumulative effect of their images creates the impression of a single panorama similar to the medieval "story" discussed earlier. The spatial coordinates of *Little Nemo*—by which I mean both the rendition of vertical events and structures within the panels, and the way that each strip as a whole demarcates a vertical shape that is usually, but not necessarily, read from top to bottom—are essential to the way in which the series holds a dialogue with the idea of temporality.

It is a matter of scholarly contention whether graphic narrative holds more temporal potential than film, because it does not hold the reader hostage to time and sequence in the same way that cinematic viewing does.[71] While the conclusions of such an argument are predicated on the specific text in question, its underlying premise—that graphic narrative is unique for its formal flexibility—indicates how questions concerning narrative sequence might be complicated by *Little Nemo*'s formal qualities. In graphic narrative, chronological coherence is fundamentally called into question by the very notion of a page of panels that can conceivably be viewed in any order, and at any speed, desired by the reader. Temporality is also specifically destabilized by *Little Nemo*'s most explicit moments of vertical precariousness, such as Nemo's repeated falls and the instability of buildings or trees. In these images of sudden and unexpected vertical movement, the subverted expectations of how action will proceed from panel to panel disrupt a sense of orderly cause and effect.

The frames in figure 6 appear to narrate a progression through time with figures ascending a staircase, but the positioning of these panels, combined with the especially narrow gutters characteristic of the *Little Nemo* series, also produces another and very different effect. Here, McCloud's suggestion that the gutter prompts the reader to create a unified reality is in some senses accurate and in others substantially lacking. The gutter does produce the overall impression that although the same place is depicted in each panel the events occur at different points in time. However, the precise placement of that gutter also makes the frames like window panes onto one single view, creating not a sense of unified reality but a nonsensical, confusing one in which the figures seem simultaneously to be at three different places in the same landscape. McCloud's and Groensteen's observations are not irrelevant to *Little Nemo*; but, even

---

71. See, for example, Chute and DeKoven, "Graphic Narrative," 770.

FIGURE 6. *Little Nemo in Slumberland*, Winsor McCay, 27 January 1907. Courtesy of Peter Maresca, Sunday Press.

when the overall impression created by a strip's panels is of consistency, there can exist a complex, and not entirely "unified," world.

At first glance, the panoramic image that straddles three panels would appear to grant a coherent reality, but, upon closer attention, it becomes apparent that this reality is fractured by the physical impossibility of Nemo existing in three different places at the same moment. If we comprehend the sequence in figure 6 cumulatively, as Ware suggests we do, the totality of the page as a building itself creates a complex world in which impossible spaces and times collide; the imbrication of *Little Nemo*'s formal construction with the representations of a vertical world within its frames creates a fractured reality that is dissonant, even unsettling. These formal disruptions are essential to the way in which the *Little Nemo* series cultivates a crossover between the action overtly described and the implied narrative of mental processes. The visual collision between immobility and movement, and the tonal collision between anxiety and pleasure, operate together to create a sense of disruption complicit with the actual vertical acts described.

McCAY'S COMICS AND UNCONTROLLABLE TIME

Scott Bukatman argues that comics provide "little utopias of disorder, provisional sites of temporary resistance"; comics and cartoons, nearly from their emergence, "set about overturning established orders and hierarchies, frequently pausing to meditate on their own possibilities." As such, Slumberland becomes a space of reverie (the daydream) "staged as a dream."[72] Slumberland produces an "aesthetic space" that is determined through McCay's construction and an "animated space" that "opens out to embrace the imaginative sensibility of a reader who is never farther than an arm's length from this other realm, a space of play and plasmatic possibility in which the stable site of reading or viewing yields to an onslaught of imaginative fantasy." Reading McCay's work in this way emphasizes the interdependency between modes of animation and the idea of plasticity at the fin de siècle. The primary claim underlying Bukatman's reading is that McCay's Slumberland is "an impermanent space"; "it's temporary."[73] Slumberland may be ostensibly temporary, but it is also not quite so. As I have discussed throughout this chapter, Nemo's states of mind and physical aspects leak over and between his

---

72. Bukatman, *Poetics of Slumberland*, 2.
73. Ibid., 1.

waking and dreaming lives, linking the two together with such intimacy that they become mutually dependent conditions.

If the action of *Little Nemo* took place exclusively in Slumberland, the line drawn between dreaming and waking life would rupture completely. *Little Nemo*'s disarmingly repetitive structure, always culminating in Nemo's return to waking life and apparent normalcy, signifies an ostensible restoration of order. However, the series destabilizes a boundary between waking and dreaming states by, in that final panel, representing dream-life spilling over into the waking world, undermining what is presumed to be discrete and echoing the way that the concept of a stream of consciousness challenged "the segmenting world view of the sciences."[74] And *Little Nemo* effects the depiction of a temporally disorienting world in more than its structural qualities. Nemo's name itself, from the Latin for "no one" or "no man," suggests a double bind in which the impossible world of dreaming is both universal and non-existent—Nemo is both not here and everywhere at the same time—and reflects back upon the confusing logic of the dream-world.

In the first strip, Nemo's fall continues as he wakes, and the sense of an accelerating fall, achieved by the conjunction of the illustrative and textual elements, is broken—but not completely dispelled—by the very final panel. Despite the relief of a return to the "real" world, Nemo finds he is indeed falling: perhaps not through limitless space, but still out of his bed onto the floor. His return to consciousness is also paired with the disjunctive return to the bright colors of his bedroom that, combined with the twisted shape of Nemo half out of bed, suggest waking as a not entirely happy jolt. Throughout *Little Nemo* and *Rarebit Fiend,* the final panel offers different configurations of this same discomfort, in falls from bed or the unsympathetic shouts of Nemo's parents. Because the final "return" to reality is an ironic covering-over analogous to the popular caveat, "it was only a dream," the final panel is a superficial return to normalcy that wryly torques the suggestion evident in later modernist texts such as T. S. Eliot's "The Waste Land" (1922) of a chaos persisting beyond the text's close. More than the mere frame to the structure of the *Little Nemo* narrative, the final panel's return to waking life ensures the preceding panels' comprehensibility. But the crossover between images and sensations—predominantly falling out of bed and a feeling of anxiety—means that Slumberland is "temporary" only in that it is briefly curtailed by the last panel. Nemo always returns to Slumberland, and,

---

74. Achilles, "Subject-Object Paradigm," 73.

more importantly, he takes his experiences and emotional registers across his journey between worlds. Ultimately, an interpretation of Slumberland as either "temporary" or not risks ignoring how the very emergence of the comic strip is a phenomenon underwritten by a convergence of conditions, innovations, and discourses related to time.

The development of photographic and cinematic technology both enabled and generated a fascination with temporal interruption and flow essential to the form of comics.[75] At the turn of the twentieth century, time affiliated with new representative technologies of photography, film, and phonography became "palpable" in an unprecedented and unique way.[76] Time was transformed into a constant visual companion through the invention of the pocket watch, underwent worldwide standardization in order to accommodate transportation and telegraph communication, and became entangled with the concept of human labor in Taylorism's measurements and Frank B. Gilbreth's "cyclographs"—photographs capturing the movement of workers as a single continuous line in space.[77] Doane argues that time became almost haptic, experienced in an entirely new manner; time was now "*felt*—as a weight, as a source of anxiety, and as an acutely pressing problem of representation."[78] The same period that saw time emerge as a calculated category—one through which a working day could be parceled out and the economic transactions of a nation scheduled according to a global marketplace—also saw Etienne Jules Marey and Eadward Muybridge producing the motion studies that are prominent in discussions of early modernism and the emergence of photographic and cinematic form. Marey's chronophotography effectively represented sequential temporal instants within one visual vector. Muybridge's series of isolated, discrete photographs captured a sequence in time while isolating each moment in the sequence. The work of both men raised questions both about the philosophic and scientific qualities of time.

As Scott Bukatman outlines, Muybridge's aesthetic offers a "fascinating disruption produced by the clearly delineated sequence of movement that demonstrated the camera's astonishing ability to register what the

---

75. For a complementary discussion of the emergence of visual media and its relation to McCay's work, in particular *Little Sammy Sneeze*, see Roeder, *Wide Awake in Slumberland*, 30–39.

76. Doane, *Emergence of Cinematic Time*, 4. See also Benjamin, "On Some Motifs in Baudelaire," 180–81.

77. Doane, *Emergence of Cinematic Time*, 6.

78. Ibid., 4 (emphasis in original).

human eye could not."⁷⁹ Such experiments with temporality, the instant, and motion were essential not only to what Doane calls the "emergence of cinematic time" but also the practices and representative modes of modernism writ large. The standardization of time—its emergence as something that could be "read" rather than merely "experienced"—was not without philosophical complications. Time's "divisibility" is hard to comprehend; "time becomes uncanny, alienated, strange." The context Doane describes is one that ruptures "time as exemplary continuum" and provides the "discursive tension" that is salient in theories of film concerned with the interplay between the cinematic frame—its moments— and its "illusion of continuous time and movement."⁸⁰ Popular contemporary accounts of cinematic technology imagined it holding "a new access to time or its 'perfect' representation," but, as Doane argues, the emergence of cinema is part of a far more pervasive mandate: namely, "the structuring of time and contingency in capitalist modernity."⁸¹

It is not coincidental that the comic strip developed a precise relationship with time at the historical moment when time itself was being rationalized. Comics emerged as a popular medium during the same period that Muybridge, Marey, and Gilbreth were experimenting in photographic motion studies; as Bukatman points out, both comics and cinema offer experiences of "temporal fracturing and temporal flow." Comics *"uniquely* present a combination of static images, often infiltrated by visual cues of captured or continuing movement, arranged in temporal sequence."⁸² Comics and cinema differ in significant ways, both in their representation of time and the way in which that representation is received by an observer. Time in comics

> is represented as territory in space, and the experience of the flow of time can be very carefully regulated, if not completely controlled. This dialectic between the stasis of an individual image and the spatiotemporal movement of the sequence—a dialectic that relates to the diegesis but also to the experience of the reader—is what McCloud calls the "temporal map," and it is a conceptual fundament of the medium.⁸³

---

79. Bukatman, *Poetics of Slumberland*, 30; see also Gunning, "Never Seen This Picture Before."
80. Doane, *Emergence of Cinematic Time*, 8; 9.
81. Ibid., 3–4.
82. Bukatman, *Poetics of Slumberland*, 31; 30 (emphasis in original).
83. Ibid., 31.

The transformation of time's representation by and through cultural events and discourses at the turn of the century is as important to the emergence of the comics form as it is to cinema. Cinema "reconstituted the movement that one could infer from the sequence of still images," whereas comics "retained the synchronous spatiotemporal array, or 'temporal map,'" but both are media that "were fundamentally bound to the explorations of time, rhythm, and tempo so characteristic of modernity." But, as various scholars suggest, the comics reader "has more control over time than the cinematic spectator, with the freedom to look back or peek forward."[84] Comics and cinema may be radically different, but their simultaneous emergence is a fact that shelters the underpinnings of both forms.

It is worth returning to Chris Ware's suggestion that comics can be read by pulling back and looking "at the composition all at once, as you would the façade of a building"; regarding the comic as "a structure that you could turn around in your mind and see all sides of at once."[85] McCay's work frequently uses the spaces of the gutter and frame to construct a world that at first glance is coherent but on closer examination presents an illogical or impossible manipulation of time, and the strips offer nonsequential reading as another way to destabilize the concept of a linear narrative. These observations might lead to the conclusion reached by Hillary Chute and Marianne DeKoven in their suggestion that, unlike the viewer of film, the reader of graphic narrative is "not ensnared in time," or that of Thomas Bredehoft, who writes that while film can "mimic some of the effects of the comics page by employing split screens and other devices," the two-dimensionality of the comics page remains essentially different from that of the cinematic screen.[86] While accepting these conclusions, I argue that it is precisely a vacillation between approximating the devices of cinema and remaining inherently static that produces *Little Nemo*'s uniquely jagged temporality.

Little Nemo frequently aligns a pervasive sense of disorienting experiences with a visual effect in which mobility and immobility are both conjured. In the strip shown below, McCay echoes the visual illusion of the stereoscope, a pair of discrete two-dimensional photographs that, positioned separately, are akin to panels of graphic narrative but, held in place, also create the visual illusion of three-dimensionality. The strip

---

84. Ibid., 31–32; 31.
85. Ware, quoted in Raeburn, *Chris Ware*, 25.
86. Chute and DeKoven, "Graphic Narrative," 770; Bredehoft, "Comics Architecture, Multidimensionality, and Time," 873; 885.

FIGURE 7. *Little Nemo in Slumberland*, Winsor McCay, 23 September 1906. Courtesy of Peter Maresca, Sunday Press.

from 23 September 1906 (figure 7) encourages a rapid reading process from left to right so that it appears as if the elephant depicted actually approaches the viewer, even if the panels ultimately remain static. Crucially, the rigidity of narrative is destabilized by a clash between apparent movement and ultimate staticity. In this sense, the "paradoxical relationship between the immobile image and the mobility of the projected film" described by Laura Marcus is complicated by *Little Nemo*'s allusions to, but resolute distinction from, other new media of the early twentieth century.[87] At the same time, in the strip from 23 September 1906, the interchange between Slumberland and waking life is expressed through a spillage of both physical and emotional sensations across sequential panels.

Morpheus's daughter, the Princess, tells Nemo that they are "a little late," and as an elephant appears to carry them to the King, Nemo cries, "I don't want to go!" The urgency of their lateness combines with the anxiety produced by the Princess's suggestion that they climb up on top of the elephant; once high up, Nemo still protests, "I don't like this one little bit, no!" Nemo never gets to experience the Princess's injunction, "Wait till you see! It's delightful up here": pleasure is terminally delayed by the real world, which breaks in to translate the dream experience into Nemo's mother saying, "Poor Mama! Was begging her little boy to get up, yes!" In both the dream and his bedroom, Nemo experiences the demands of urgency (to get to King Morpheus; to wake up) and the physical experience or anticipation of "up" (getting on top of the elephant; getting up out of bed). Once more the vacillation between dreaming and waking gives the "reality" of the final panel a tenuous quality. While this vacillation is characteristic of each strip in the *Little Nemo* series, figure 7 is particularly significant for its emphasis on "up" as the vehicle for the experience of anxious sensations that bridge the real and dreaming worlds.

The tension between the motion depicted and the ultimate immobility of the drawings is joined with the conflicts generated between Nemo's simultaneous fear of, and complicity in, climbing on top of the elephant. Nemo acquiesces to a foreboding vertical act, despite, or perhaps because of, his fear. As in the strip above, throughout *Little Nemo,* verticality is relentlessly accompanied by Nemo's "intense anguish" as he feels himself fall through limitless space, has objects fall upon him, is threatened by impalement or drowning, and experiences the fragility of the surface upon which he is standing. In the strip from 22 October 1905, Nemo

---

87. Marcus, "'New Form of True Beauty,'" 269.

falls through his bedroom floor into an underworld of "tender" (fragile) over-sized mushrooms only to become fatigued and touch one mushroom, so that all the mushrooms then fall down upon him like dominos until he awakens, terrified and literally falling out of bed. The overlap between Nemo's physical sensations in his waking and his sleeping life—in the real world Nemo is tired and falls asleep; in his dream he is still tired, and it is this tiredness that causes him to knock over the first mushroom—creates a loop so that the physical discomforts of both worlds are not distinct. Here and elsewhere, Nemo's dream transforms into a distressing experience; the reimposition of real life cannot quite secure him from the uncontrolled dream-world.

The way that such dissonant visual practices unravel the idea of a coherent or stable temporality is best understood in terms of an "interplay between the still and moving image," and what this interplay suggests about time.[88] Laura Marcus reveals that the "issues central to an understanding of cinema (including questions of time, repetition, movement, emotion, vision, sound, and silence)" are "threaded through" the modernist writing that fictionally or discursively engaged with the impact of film.[89] Some modernist writers "used their writing as a way of halting, or attempting to halt, the flow of film in an effort to capture cinematic essence," whereas others wanted less to "crystallize or to conquer the fleeting image or instant" than to "trace the lineaments of motion itself." The new cinematic forms that evolved into the avant-garde and experimental film of high modernism were extensions of the central tension between stasis and mobility; the suggestion that film could transform the mechanical into the living became the underlying preoccupation of discourse surrounding film.[90] If film was enticing precisely for its seeming approximation of real movement, the comic strip both participates in this simulation and crucially withholds its complete realization; to some extent, McCay's dream-comics strive to reproduce the movement that defines cinema, a movement that "(as it comes so often to define modernity itself) is also a more fragile and unstable ephemerality" that "inheres at every level," from "the fleeting nature of the projected images" to "cinematic exhibition."[91]

Even when McCay's work seems to project a unified reality, it is subject to the crossover of characters between the apparently discrete worlds,

---

88. Ibid.
89. Idem, *Tenth Muse*, 2.
90. Idem, "'New Form of True Beauty,'" 269.
91. Idem, *Tenth Muse*, 2.

such as that of *Jungle Imps* and of *Little Nemo,* as well as its more localized leakage within individual strips, where the sensations that occur in waking life and dreaming permeate one another. While these characteristics make certain destabilizing tendencies part of the narrative of *Little Nemo,* the tension between stasis and motion I have discussed in reference to figure 7 further erodes the idea of integrity by calling into question the disparity between stillness and the appearance of movement. Crucial to this is the fact that, in graphic narrative, the images are not merely "illustrative" of the text, but, in fact, move "forward in time in a different way than the prose text"; the visual and verbal cues of graphic narrative do not create a "unified whole," but rather "remain distinct."[92] Throughout the dream-comics, it is the dissonance between word and image of which Chute et al. speak that allows the series to exploit the conversion of dream content from visual to verbal as a disorienting one.[93] While critics agree that the mixture of word and image lends graphic narrative greater combined weight, they remain divided as to whether the bifurcation created by this combination ultimately prioritizes one of the two elements over the other.[94] I propose that the dream-comics actively cultivate the distinction itself in order to procure a fragmented progression of meaning. As such, McCay's work maintains the staccato rhythm produced by a contrast between textual and visual information while simultaneously summoning the fluid temporality associated with film.

Walter Benjamin wrote that the mechanical reproduction of the camera "introduces us to the unconscious optics as does psychoanalysis to unconscious impulses." He argued that, "with the close-up," "space expands; with slow motion, movement is extended"; the "enlargement of a snapshot does not simply render more precise what in any case was visible, though unclear: it reveals entirely new structural formations of the subject."[95] Benjamin's "unconscious optics" of cinema are of equal significance to *Little Nemo,* a text that both summons and holds at bay the motion inherent to film. While an abundance of writing that attaches psychoanalysis to cinema argues that film is the closest relative

---

92. Chute and DeKoven, "Graphic Narrative," 769.
93. See Hirsch, "Collateral Damage"; Chute, "Comics as Literature?," 452.
94. For a clear example of these two opinions, compare McCloud's *Understanding Comics* with Groensteen's *System of Comics.* The former suggests an overall coherence is achieved, whereas the latter contends the supposed cohesion of word and image, arguing for the "primacy of the image" (3; 8). Chute and DeKoven more closely approximate my stance when they argue that word and image do not "merely synthesize" ("Graphic Narrative," 769).
95. Benjamin, *Illuminations,* 229–30.

of dream-interpretation because "while the dream disguises unconscious wishes and desires as a way of eluding the censor, the film reveals them," graphic narrative is at least as close to approximated psychoanalytic procedures.[96] In this sense, the way in which *Little Nemo* is structured—episodic and sequential but also holding within itself an immanent potential for nonlinear reading—dramatizes the question of how consciousness can be represented by graphic narrative in a necessarily different, but no less potent, way from literature or film.

Little Nemo might be understood as straddling several points along a continuum of visual representation that comprises the most basic static drawing, the illusion of movement in sequential narrative and the stereoscope, and the three-dimensionality represented in cinema. Doane has interrogated how "the cinematic image as a representation *of* time" intersects with the philosophical construction of the "event" as a temporal idea fraught with contradictions and paradoxes.[97] The emergence of cinema as a narrative form at the beginning of the twentieth century posed a set of questions about selection and capture, and about contingency and structure. McCay was one of the earliest animators, and it is not incidental that animation problematizes a similar set of discourses discussed by Doane. Animation, after all, bears the same formal quality inherent to cinema: it is composed of individual static frames—"'instants' of time"—that, when played sequentially, create an illusion of movement.[98] But animation is structurally different from cinema; it is always selective—the effort required to devise each frame means that McCay's early animated films could not be thought to capture chance in the way that the actuality films did. Moreover, animation "mimics, parodies, the cinema": the still frame of a cartoon "is not a frozen moment of movement but rather the basic unit of the animated film: the single drawing."[99] The distinction is important because it indicates how the comic strip and its moving sibling, the cartoon, are radically different from the photograph and its moving sibling, cinematic film.

Gilles Deleuze writes that, if the cartoon film "belongs fully to the cinema," it is because "the drawing no longer constitutes a pose or a completed figure, but the description of a figure which is always in the process of being formed or dissolving through the movement of lines and

---

96. Marcus, "Cinema and Psychoanalysis," 244.
97. Doane, *Emergence of Cinematic Time*, 141 (emphasis in original).
98. Ibid., 9.
99. Bukatman, *Poetics of Slumberland*, 45.

points taken at any-instant-whatevers of their course."[100] In strips such as the one shown in figure 7, *Little Nemo* both maintains the distinction between panels and equally strives to create the effects of time lapse animation McCay would experiment with in later years. The comparatively narrow gutter characteristic of *Little Nemo* makes the procession from left to right a fluent process, but the images within each panel remain distinct so that, lacking the simultaneity of viewing two stereoscopic images, the strip is ultimately unable to approximate the dynamics of film or even fully create the effect of a stereoscope. Nonetheless, the series generates the same question of how to represent movement and time that is implicit in animation, and uncovers how this question is interdependent with psychoanalytic discourses.

Verticality in the dream-comics typifies a fairly straightforward relationship between vertical imagery and feelings of epistemological discontinuity engendered by revolutions in psychology and psychoanalysis. McCay's work also captures the anxiety of racial and cognitive difference, but the dream-comics do not attempt to work through these anxieties in any meaningful way. In fact, *Little Nemo* and *Rarebit Fiend* are formally and thematically trapped in their own repetitiveness—a characteristic that ensures they reflect more completely the idea of psychic repetition with which they are preoccupied, but does not imbue vertical imagery with the power to intervene in the social and historical conditions its narrative relates. The mediation of vertical occasions and experiences through a collision of simulated motion and the static image allows *Little Nemo*'s aesthetic to approximate disjunctive psychic processes. At the same time, the comprehension of word and image in *Little Nemo* replicates a process of translation just as dream interpretation came to represent a form of auto- or biographical translation. These formal conditions articulate the general sense of dissonance effected by contemporary models of consciousness. *Little Nemo* marks the beginning of tropes that reverberate through the twentieth century in conjunction with scenes of verticality and, in its representative methods, makes the disorienting power of new ideas about what it meant to be conscious into a direct correlative to the disorienting experiences of its protagonist. If Nemo is not explicitly waiting for the sky to fall, the temporality of the comic strip and the more widely communicated sense of anticipation he experiences in it produce a verticality determined by the rigors of time.

---

100. Deleuze, *Cinema 1*, 5.

CHAPTER 2

UPTON SINCLAIR'S VERTICAL INFERNOS

Oil Procurement and Disaster Culture

DISASTER CULTURE, VERTICAL CATASTROPHE, AND OIL NARRATIVES

Oil's refinement into fuel is the story of a peculiarly democratic process, repetitive moments of disaster, and overwhelming spectacularity. First: the rule of oil capture allowed for the ownership of an oil well to follow from discovery of oil rather than land rights; if you found oil, you could claim it, regardless of location. Second: the process of establishing oil wells precipitated a series of risks based on the volatility of oil and its flammable nature, as well as the expected increase in danger to human life as a result of accidents that took place during drilling. Third: the erection of towering derricks, the impressive spurt of oil from the ground, and the visual impact of an oil well caught on fire promises a set of stimulations that, in Stephanie LeMenager's words, are "visual, kinesthetic, acoustic (hissing), tactile, olfactory."[1] Although the specifics of oil

---

1. LeMenager, *Living Oil*, 101.

drilling disasters have changed, the catastrophic event itself remains an accepted risk of the process of drilling.

Oil procurement, rendered into its barest aesthetic terms, comprises a triple verticality: the descent of machinery into the earth, the erection of derricks and buildings surrounding the drilling site, and the occasional but spectacular gush of oil spurting into the sky from a break in the ground. Oil's verticality is, moreover, intimate with the kind of aspirational sentiment lodged in texts representing the skyscraper and space flight. As LeMenager notes, "ultradeep" drilling, which involves descending 5,000 feet or more, is "an extension of the once space-age ambition of deepwater drilling: to go as far as 1,000 feet."[2] The affiliation of space-age ambition with deepwater drilling is not, here, a specific comment on the relationship between the two. Nonetheless, as Michael K. Walonen discusses, oil's ability to "safely and efficiently power the internal combustion engine" radically altered human social, economic, and spatial conditions; there is an explicit link between the engine and the emergence of atomic technology, modern warfare, and space exploration.[3] As such, both the ontological properties of oil and the various textual representations of its procurement refer to a set of ideas that pervade the technological feats of the atomic bomb and space flights. These are subjects I reprise in chapters 3 and 4 of this book.

This chapter focuses on the histories and narratives of producing oil, tracing the industry's development in the first quarter of the twentieth century as a dominating economic force in American culture. It explores how Upton Sinclair's 1927 novel *Oil!* simultaneously extols and cautions against the economic success ensured by the oil industry. Sinclair's representation of the vertical oil geyser and oil rig fire is indicative of wider conflicts between ideas of financial prosperity and human risk in discourses of the 1920s concerning the oil industry. The following argues that Sinclair's work bears a set of representations and narrative influences that have caused it to be overlooked in scholarship. Calling this form *melodramatic didacticism*, I contend that the frequently overblown, exaggerated, and hyperbolic prose of *Oil!* is intimately related to the contemporary crisis of understanding around the unknown consequences of the nascent oil industry. Ideas of danger, payoff, and uncertainty percolate through the vertical imagery of *Oil!* to reveal how representing vertical disaster in American texts becomes complicit with questions of what qualifies as acceptable risk.

---

2. Ibid., 3.
3. Walonen, "'The Black and Cruel Demon,'" 56.

On 26 December 2010, the *New York Times* published an article entitled "*Deepwater Horizon*'s Final Hours." In this article, David Barstow, David Rohde, and Stephanie Saul describe the explosion of the *Deepwater Horizon* oil rig in the Gulf of Mexico in April 2010 as a scene in which crew members were "hurled across rooms" and "buried under smoking wreckage." In their words, the oil rig was transformed into a "firestorm" of "searing heat" that forced men into the "oily seas 60 feet below." The extravagant imagery adopted by the *New York Times* typifies media coverage of both the explosion itself and its catastrophic environmental repercussions along the Louisiana coast, which continued to dominate the news for a large part of the year. While depictions of the *Deepwater* fire reveal the degree to which apocalyptic narratives engage the public imagination, the disaster also ensured a greater audience for several documentaries already made or released when the rig exploded. *Blood and Oil* (2008), *Dirty Oil* (2009), *H2Oil* (2009), and *Petropolis: Aerial Perspectives on the Alberta Tar Sands* (2009) all scrutinize international politics and the environmental effects of oil dependence to imply, like depictions of *Deepwater,* that catastrophe is an inevitable consequence of the oil industry.

The iterative nature of oil disaster, in LeMenager's words, ensures that "every oil spill remembers every other, from the mid-twentieth century onward"; but every time there is an oil disaster, it is as if the event is "happening for the first time." LeMenager analyzes the "intimate miseries of oil spills" in terms of what Derrida would call "the poetic and therefore untranslatable, untransmissible quality of witness testimony," which "meets extra resistance from complicit cultural memory, a national imaginary saturated in oil."[4] LeMenager's critique of petroleum culture—namely, the "objects derived from petroleum that mediate our relationship, as humans, to other humans, to other life, and to things"—marks the oil disaster as a moment that is both omnipresent and repetitive. The pervasiveness of petroleum culture—its entrance into every crevice of contemporary American life—secures a condition of being "in oil, living within oil, breathing it and registering it with our senses."[5] If oil is by now so ubiquitous that we both live oil and its disastrous consequences, volatile scenes of oil extraction underpin narrative representations of the industry since Upton Sinclair's largely overlooked novel, *Oil!* (1927). This chapter reads texts that represent subsequent disaster as a seeming

---

4. LeMenager, *Living Oil,* 64–65.
5. Ibid., 6.

fulfillment of oil's volatility. In Sinclair's novel, the vertical extraction of natural resources from the earth both harbors an immanent threat and also produces a rapture via the spectacular fulfillment of that same danger. The novel produces a relationship between oil drilling and an impending disaster that seems always ready to transform the extraction scene into a vertical inferno, a relationship reminiscent of that between the skyscraper and its vertical collapse I discussed in chapter 1.

Sinclair is best remembered for his polemical novel *The Jungle* (1906), which exposed the horrific working conditions of the meatpacking industry and established its author as a passionate critic of American capitalism. Sinclair has received comparatively scant attention for the volume of prose that he produced, although his colorful personal life has ensured several detailed biographies.[6] *Oil!* is a sprawling and didactic novel in which initial and highly dramatic images of oil drilling give way to an extended political and social critique; sensational in both tone and content, the novel frequently breaks into propaganda and a theatricality reminiscent of the dime novels Sinclair started his career by writing. Studies exploring Sinclair's fiction have tended to emphasize these characteristics, forging the general opinion that his work is more noteworthy for its attempt to encourage social reform than for its literary weight; that it lacks in finesse, has a melodramatic tenor, and sacrifices narrative depth for political rhetoric.[7] Recent critical examinations of Sinclair's work, however, have started to countermand the earlier emphasis on his work as being merely social rhetoric.[8] I use the term *melodramatic didacticism* to refer to both the novel's sensationalist tone and the underlying complexity that is occasionally occluded by this sensationalism. *Oil!*'s evocation of competing or apparently incompatible discourses—capitalism and socialism, pleasure and anxiety, disaster and the sublime—coalesces around pervasive imagery that emphasizes the verticality of oil procurement, whether in the form of the drill plunging into the earth or the towering derrick and the flames that seem constantly ready to erupt from it.

Sinclair's description of an oil well fire is, in fact, uncannily similar to the 2010 description of the *Deepwater* explosion I discussed earlier.

---

6. Harris, *Upton Sinclair*; Yoder, *Upton Sinclair*; Bloodworth, *Upton Sinclair*; Arthur, *Radical Innocent*.

7. See, in particular, Mookerjee, *Art for Social Justice*.

8. For example, the papers included in Herms, ed., *Upton Sinclair*, approach Sinclair's work within a theoretical framework of a proletarian aesthetic, Fredric Jameson's political unconscious, and feminist models. See Riese, "Upton Sinclair's Contribution"; Hornung, "Literary Conventions"; Kerkhoff, "Wives, Blue Blood Ladies."

The novel's protagonist, Bunny Ross, watches the drilling at his first well when he experiences a sound that is "literally a blow on the side of his head"; suddenly "an express train" of water and oil is "shooting out" of the earth. The "black floods" that are propelled into the sky then ignite; there follows "a tower of flame, and the most amazing spectacle" that is depicted in appropriately breathless tones:

> The burning oil would hit the ground, and bounce up, and explode, and leap again and fall again, and great red masses of flame would unfold, and burst, and yield black masses of smoke, and these in turn red. Mountains of smoke rose to the sky, and mountains of flames came seething down to the earth; every jet that struck the ground turned into a volcano, and rose again, higher than before; the whole mass, boiling and bursting, became a river of fire, a lava flood that went streaming down the valley, turning everything it touched into flame, then swallowing it up and hiding the flames in a cloud of smoke.

The "mountains" and "jets" of lava that rise high into the air are a cataclysmic vertical event. The valleys are overrun by a "river" of fire and a "flood" of lava, signaling an inexorable and consumptive force that is hungrily "swallowing" up the terrain. Exceptional for their sheer length alone, the two sentences portraying the fire syntactically accumulate force; each successive clause added to the description effects a sense of relentlessness that mirrors the uncontrollability of the blaze and the image of the smoke that rolls forth from it. The description is more than merely extravagant: it is overtly apocalyptic, and the prose suitably anxious. Yet the language is also excited; the prose exclaims, "You saw the skeleton of the derrick, draped with fire!"; Bunny registers the fire as a "dreadful thing," but is also swept away by the "amazing spectacle" before his eyes.[9] The scene is exemplary of the way that verticality becomes complicit with the problems of oil procurement in *Oil!*.

Vertical catastrophe is made visually explicit by Paul Thomas Anderson's *There Will Be Blood* (2007), the film inspired by Sinclair's novel. Anderson's film transcribes Sinclair's "tower of flame" into a symbol of the costs of consumer culture. At a crucial juncture in the film, the oil tycoon Daniel Plainview stands in the foreground as one of his derricks burns against a night sky; in the scenes immediately before, the process

---

9. Sinclair, *Oil!*, 160–61. All subsequent citations come in the main text in parentheses.

of drilling has released a volatile stream of oil and air into the sky only for this geyser to ignite and become a blazing inferno. As in Sinclair's novel, the fire suggests that the derrick rests upon an always potentially incendiary activity. And, like Sinclair's novel, the film is equally notable for its simultaneous rapture and disgust in the face of what oil drilling implies. *There Will Be Blood* portrays Daniel Plainview as the mythological self-made American man who goes from silver miner to oil man, but the protagonist's ascent to successful millionaire is skewed by the film's ending, which depicts Plainview beating the young evangelical preacher Eli Sunday to death. *Oil!*'s J. Arnold Ross, the inspiration for Plainview, has an even more dramatic transformation from mule-team driver to oil millionaire, and the novel also withholds a triumphant depiction of Ross's accumulation of wealth. Although Ross is described as "a typical American, risen from the ranks, glorifying once more this land of opportunity," the novel is simultaneously exhilarated and concerned by the cost associated with such a "rise" (34). Throughout, Ross is portrayed bearing an "ugly determination" and a desire to make "a tremendous killing" in the oil market (11; 248).

Contemporary texts concerning oil extraction capture a trade-off between, on the one hand, wealth and, on the other, massive human and environmental cost. *There Will Be Blood* returns to the fin de siècle to narrate a morality tale made timely by its study of, in Kenneth Turan's description for the *Los Angeles Times* on 26 December 2007, the "rapacious, uncaring capitalism" of modern American culture. In the film, a retroactive return to the early twentieth century compounds the prediction laid out by *Oil!* that a desire for oil products would dictate more than the economic history of the twentieth century. While oil, capitalism, and war are thematically joined in Sinclair's novel, *There Will Be Blood* transforms the oil well fire into a coherent symbol for an audience familiar with oil disasters such as the *Exxon Valdez* spill in 1989 and the Gulf War spill in 1991 and, in its contemporary context shaped by American military efforts in Afghanistan and Iraq in the 1990s and after 9/11, places war as a consequence of the desire for a secure source of oil.[10] The conflagration is a shorthand for the film's own internal narrative arc; the fire is bound together with Plainview's volatile temper so that the inferno operates within and without the film as a symbol of the implied

---

10. Commentators such as Mark Lawson were quick to point out the resonance of *There Will Be Blood* in light of Iraq and Afghanistan. See *The Guardian*, 16 February 2008, "'The Land of Hope.'"

costs—both literal fire and human wrath—of consumerism. As such, the film adaptation reconfigures the contexts of Sinclair's novel, flattening out the disparity between historical moments almost a century apart to reflect upon the pervasiveness of those anxieties attached to an industrial progress that is only ostensibly positive.

While *There Will Be Blood* ends in explicit violence, Sinclair's novel suggests oil's unsavory consequence more obliquely. Bunny asserts that his father is "sly as the devil" and that the oil industry is a "dirty game" (299). However, while less explicitly violent than his filmic counterpart, Ross is nonetheless depicted as corrupt, more concerned with using his "twenty million" dollars to buy the support of the President than ensure the welfare of his workers. As his workers become more impoverished, Ross considers buying a tract of government land that has been reserved for the Navy in order to secure the wealth of oil lying beneath it (296–97), echoing a real political scandal in 1922 concerning the Harding Administration, in which the Secretary of the Interior covertly leased a sum total of 67,000 acres of oil reserves in California that had been set aside for the Navy.[11] Ultimately, both Plainview's physical violence in *There Will Be Blood* and Ross's self-interest in *Oil!* compromise the positive American myth of self-made wealth by suggesting the human costs of capitalism.

The rendition of the oil well fire in Anderson's film chimes with the general tenor of Sinclair's writing, which consistently concerns itself with the consequences of an emerging capitalist society. Sinclair's fidelity to the history of oil drilling also makes the novel operate retrospectively as a genesis story: a narrative of origins which, for a contemporary audience, justifies the ongoing significance of the anxieties surrounding oil extraction. Whereas the novel's early stages—in which Bunny and his father pursue the oil underneath the Watkins family ranch—seem to endorse Bunny's "dream" of discovering oil himself as a viable and even admirable fantasy, it quickly becomes clear that the relationship between oil drilling and war taints Bunny's optimistic fantasy of "digging a hole in the ground" and seeing the oil simply "come spouting up" (97). The unsavory ethical problems associated with oil drilling come so quickly after Bunny's dream of discovery that it seems one is the direct result of the other. The oil Bunny knows to lie under the Watkins ranch throws into relief a "moral problem" that will shape Bunny's relationship with Paul and Ruth, the children of the Watkins family, and will undergird the narrative's consideration of morality and industry. Bunny wonders:

---

11. See Scott, *Upton Sinclair*, 256–57.

"Just what rights did the Watkinses have to the oil that lay underneath this ranch?" (98). It becomes clear to him that in order to be successful he must deny the Watkinses the wealth lying below their home; oil drilling signifies the human brutality that capitalist success seems to necessitate.

This chapter argues that, while mediating the ideas surrounding specifically new vertical endeavors of drilling and mining, Sinclair's novel characterizes a particularly American rapture with scenes of overwhelming scale and grandeur. At the same time, the novel complicates this same rapture with a critique of capitalism. The novel's depictions of oil gushing from the earth and the overwhelming height of the vertical derrick suggest violence, masculinity, and industry; the downward thrust of the oil drill into the earth implies another kind of violence to do with the human cost of capitalism. The novel's ambivalent tone and various narrative voices make the text a clash of discourses that, in aggregate, refuse a depiction of oil drilling as unequivocally desirable or entirely irredeemable. This lack of resolution makes vertical imagery a series of interrogative gestures that question why precisely oil procurement is so welcomed by American culture.

IMAGINING HORIZONTAL AND VERTICAL SPACE

Sinclair's depiction of the oil well fire accords with a long-standing tradition of imagining California as an inherently precarious landscape. Variously discussed as the ultimate American frontier and a cauldron for anxieties surrounding natural disaster, Southern California is frequently the stage for fictional chaos; as Mike Davis famously suggested, nothing seems to "excite such dark rapture" as that provided by the image of Los Angeles tumbling into the Pacific or being "swallowed up" by the San Andreas fault.[12] As both a last frontier and the stage for an imagined apocalypse, California is an extremity: at the very edge of the map and also perpetually on the cusp of a natural disaster.[13] In this sense, California's horizontal spatiality creates similar narrative associations as those offered by Manhattan, which, as I described in the introduction to this book, is frequently the location for vertical imagery precisely because it is horizontally circumscribed by an inflexible shoreline. Here, California's

---

12. Davis, "Dark Raptures," 8. See also idem, "Isle of California"; *City of Quartz*; *Ecology of Fear*.

13. See Hazard, *Frontier in American Literature*; Haslam, "Literary California"; Durham and Jones, eds., *Frontier in American Literature*.

location at the horizontal edge of the national map intensifies the vertical imagery used in its representation. California, then, is already imbued with particular resonance as a volatile locale. Sinclair's depiction of the oil well fire marries extremity of place (California at the edge of the map) with extremity of event (potential apocalypse) in a specific and extravagant combination of horizontal and vertical space.

While Sinclair's "tower of flame" specifically performs the prophesied Californian apocalypse, it also metonymically gestures toward the problems associated with oil use. As Paul Sabin argues, California provides a microcosm of the "penetration" of "oil" throughout the United States, exemplifying both the national dependence on car travel and the subsequent high levels of pollution caused by traffic.[14] California also betrays the immediate effects of oil consumption in scenes of what Nancy Quam-Wickham characterizes as "pollution, overproduction, and profligate waste."[15] Working from his own eleven-year residence in Pasadena, Sinclair depicted the Californian oil industry as reflecting these historical trends. The fictional "Prospect Hill field," site of the "greatest oil strike in the history of Southern California," is a facsimile of Signal Hill, one of the most prosperous fields in the region (25). Sinclair's rendition of oil extraction is equally factual; Californian oil fields were subject to regular and spectacular "blow-outs, gassers, gushers, fires, explosions, craters, geysers and other exciting incidents."[16] Infernos were also common, tallying into the hundreds by the end of the 1920s and remaining a "vivid and enduring image" for those who spent their childhoods close to oil fields. In her description of the "unchecked oil development" that meant oil "flowed relentlessly and uncontrollably from Southern California," and left the entire industry "on the verge of collapse," Quam-Wickham uses adjectives and imagery that render the oil industry on the precipice of its own destruction.[17] *Oil!*'s depiction of a "swallowing" river of flames that seems always ready to usher in the apocalypse bears the same semantic weight; the novel forges a link between, on one hand, the very real ruin ensured by the seemingly unstoppable aftermath of oil consumption and, on the other, the always impending end attached to the Californian landscape.

The apocalyptic trope rendered in a vertical symbol of disaster is coupled in *Oil!* with the pervasive horizontal motion of the automobile in the novel's extended opening. In the introduction, I argued that a focus on

---

14. Sabin, *Crude Politics*, 1–11.
15. Quam-Wickham, "'Cities Sacrificed,'" 190.
16. H. C. George in the popular press, quoted in ibid., 192.
17. Ibid., 193; 190–91.

horizontal motion before 1900 reflects the significance of the frontier as a primary conceptualization of space in America and that, in texts after the frontier's closure, a conspicuous emphasis on vertical direction emerges. Sinclair's use of the vertical frontier as a uniting metaphor is especially complex. While the insistency, violence, and complex tenor of images in which oil bursts from the earth ultimately override a horizontal topography with a vertical one, *Oil!* begins with a horizontal car journey.[18] Yet even the ostensibly horizontal early narrative action situated within the car alludes to the later verticality of the plot; the novel's shift from horizontal to vertical imagery is calibrated by a transition from movement in the automobile to the more powerfully depicted vertical movement of oil geysers and drills. Crucially, the novel's opening depicts Bunny Ross and his father traveling in a generally horizontal direction over the California roads, while at the same time evoking the vertical wavering of that same road in its ascent and descent over mountains.

Verticality is also the subtext in allusions to the car's "oil gauge" and "gas gauge," the "filling station" at which the family stops, and the description of a "softly purring engine running in a bath of boiling oil" (7; 11; 14), all of which annex the car's horizontal journey to the oil drilling needed for its successful completion and prefigure the drilling depicted by later parts of the novel. As such, the novel's opening constructs the image of the road as the convergence between verticality and capitalism; literal vertical motion and the implied verticality of drilling coalesce through their shared relationship to petroleum. The depiction of a car moving quickly over the road also solicits the particular visuality provided by the automobile, an experience Robert Hughes describes as the compression of "horizontal visual information" into "a new experience of space."[19] At the same time, the speed of Ross's car recalls the pace with which the automobile was being produced in the new industrial technique of the assembly line. The assembly line emerged as the culmination of the subdivision of labor in Henry Ford's Highland Park factory in Detroit and meant that, after 1910, the automobile was constructed at both a greater speed and a lower cost. David E. Nye calls the method a perfect manifestation of the "acceleration" of experience born from "the American fascination with speed."[20] The literary result in *Oil!* is a scene

---

18. See also LeMenager, *Living Oil*, 69–70; 80–81, for a reading of *Oil!*'s opening that focuses on the "persistent association of driving with being alive" (80).

19. Hughes, *Shock of the New*, 12. See also Nye, *American Technological Sublime*, 45–76; 75; idem, *America as Second Creation*, 147–73.

20. Idem, "Accelerating American Experience," 1.

that reacts to the swiftness of industrial change by making that rapidity manifest in the immediate physical speed of a car journey.

While evoking the rush of technological advance, the car journey in *Oil!* also suggests the potential danger associated with speed as Bunny and his father continually risk collisions with other drivers (3–7). Oil and the automobile it feeds serve a connected function as symbols of industrial progress, united by their reliance upon heavy industry but also compromised by the underlying suggestion of danger in the narrative. This danger leaks into a scene that nonetheless seems to hold onto the fundamental beauty of the landscape observed from the car, revealing a tension in the novel between the natural world and a new industrial one. Sinclair frequently extolled the power of a natural landscape as a haven from society and a source of inspiration; he is particularly ebullient when writing about his solitary encounters with wildlife. In his autobiography, he declares that "tree branches white with snow, magical in the moonlight" and valleys "full of clover" are the preeminent source of inspiration; in 1903 the author even moved his family to a tent in the New Jersey woods so that he could spend approximately sixteen hours a day writing in solitude.[21] In some ways, Sinclair's automobile proceeds from the tradition of the unwelcome machine intruding upon nature described by Leo Marx, in which the disruptive agent is frequently a locomotive bursting onto the tranquility of a natural scene. Nominated by Marx as the time when the balance between the agrarian tradition and the burgeoning industrial system had the "greatest literary impact," the reigning trope in these narratives is that of an "interrupted idyll," and the underlying sentiment is an understanding of the car or train as anathema to a natural, and peaceful, world.[22] However, Sinclair's imagery is not as unilateral as Marx's model: it remains unclear if the automobile and the road, as twin emblems of industrialization in *Oil!*, are welcomed or critiqued. On one level, the enjoyment Bunny garners from the car journey implies that American industry holds an allure equal to that of the natural world, suggesting that Sinclair's depiction of the car is a riposte to the more traditional construction of machinery breaking into and disrupting a natural scene. On another level, even in these terms *Oil!* vacillates: it is neither exclusively exhilarated by, nor consummately anxious about, what the interruptive car signifies, but rather grasps for both feelings at the same time.

21. Sinclair, *Candid Reminiscences*, 73ff.; Bloodworth, *Upton Sinclair*, 40–41.
22. Marx, *Machine in the Garden*, 26ff.

The novel's emphasis on the new vertical expansion in America is a reflection on the swift technological and economic changes following the Civil War that made the verticality of a new landscape inextricable from the industry that had enabled it. Alan Trachtenberg suggests that texts contemporaneous with *Oil!* alternate between attaching either "plenitude" or "devastation" to this industrial progress.[23] However, Sinclair's novel does not alternate between, but rather marries, the polarities invoked by other texts. Moving from horizontality to verticality in order to reflect new ways of conceptualizing space, *Oil!* registers a conflation of apparent opposites through the disparate tones of its prose. In the depiction of the road, the dual tonality endemic to the novel as a whole emerges in the movement between economical language and a more poetic phrasing:

> The road ran, smooth and flawless, precisely fourteen feet wide, the edges trimmed as if by shears, a ribbon of grey concrete, rolled over the valley by a giant hand. The ground went in long waves, a slow ascent and then a sudden dip; you climbed, and went swiftly over—but you had no fear, for you knew the magic ribbon would be there. (1)

The adjective used to describe the road here—"trimmed"—has a blunt phonetic quality that gives it an almost onomatopoeic resonance; the noun "concrete" parallels the descriptive brevity of "trimmed," while the phrase "precisely fourteen feet wide" adds banal, exact measurement to the prose. However, with the entrance of verticality onto the scene, the narrative betrays a breathless, lyrical tone; the extended phonemes of "rolled" and "slow" protract the bluntness evoked by the terms "trimmed" and "concrete." The pace of the language is carefully measured in order to mimic the motion felt in the car. The measured first sentence gives way to the more hurried second one, which then breaks at the point of declaring, "You . . . went swiftly over—but you had no fear." The prose continues to describe the ascent and descent of hills: "The car raced on—up, up, up—until it topped the ridge, and was sailing down into the next little valley" (2). Going "over" the crest of a hill and rushing "up, up, up"—a direction so crucial it is repeated three times—transforms the prosaic beginning of the passage into a "magic" experience that hinges upon verticality. The narrative voice grows reverential as it depicts the "barrier of mountains" that lie "across the road"; they are "far off" and

---

23. Trachtenberg, *Incorporation of America*, 52.

"blue, with a canopy of fog on top"; they lie "in tumbled masses, one summit behind the other, and more summits peeking over, fainter in color, and mysterious." The romanticism of the prose finally takes on a compulsive intonation as it asserts, "You knew you had to go up there, and it was interesting to guess where a road might break in" (4); already vertical ascent is more intriguing than horizontal motion and, notably, is also the source of a seemingly hypnotic power.

In *Oil!*, the doubled invocation of excitement and anxiety is a deeply entrenched conflict, made most explicit by the novel's ambiguous narrative voice. This voice both satirizes and is affectionate toward Bunny's naïveté and, when revealing a crisis of confidence in the natural world discussed earlier, prefigures the ambivalence surrounding vertical imagery in the later parts of the novel. In particular, the narrative voice's indefinite tone while describing Bunny's car journey is paralleled by its depiction of oil drilling, which, as a widely acknowledged symbol of industrial progress, shares the potential rupturing influence of the steam train bursting onto a pastoral scene as well as engendering a comparable polarized public reaction. In the novel's opening, the narrative voice approximates the novel's protagonist while also projecting a satirical and more worldly undercurrent. The direct address of "You knew" implies a direct commune between Bunny and the reader, and "it was interesting to guess where a road might break in" suggests the kind of naive wonder usually apportioned to the boy in the early stages of the novel. At the same time, the description of "the boy," "eager-eyed, alert," implies a suspicion of his fantasizing, through which he is "seeing the world" in "a fashion men had dreamed in the days of Haroun al Raschid—from a magic horse that galloped on top of the clouds, from a magic carpet that went sailing through the air" (6). The idea of the car journey as a magic carpet ride from an Arabian fantasy foreshadows Bunny's later belief that the Ross-Junior oil well will "yield him a treasure that would make all the old-time fairy tales and Arabian Nights adventures seem childish things" (160). Here, his imagination transfigures the road, making it "a giant's panorama unrolling itself; new vistas opening at every turn, valleys curving below you, hilltops rising above you, processions of ranges, far as your eye could reach" (6). But his fairy-tale rapture collides with the simultaneous implication that industrialization augurs destructive results. Here and elsewhere, the association of vertical drilling with a prevalent sense of anxiety matches the imagining of California as both physically at the edge of a horizontal map and metaphorically on the precipice of disaster, thereby dramatizing a broader leakage between physical space and the abstracted ideas attached to it.

While the novel summons the volatility of the Californian landscape and its propensity for volcano-like eruptions, earthquakes, and geysers, it also frequently uses verticality to condense a relationship between imminent jeopardy and aesthetic pleasure. Like in the description of the car journey, while describing Bunny watching the drilling process at Prospect Hill, the prose flirts with the idea of danger while simultaneously seeming to relish the sight of the "lovely dripping black" oil. The narrative voice declares, "You understand, the pressure of the gas and oil was held down by the column of water, two thirds of a mile deep," then continues to describe the process of lifting out "fifty feet of the water-column" to "dump it into the sump-hole" as pressure builds (77). Finally, "the water would be shot out of the hole, and mud and water and oil would spout up over the top of the derrick, staining it a lovely dripping black. You must drive the crowds off the lease now, and shout 'Lights out!' to the fools with cigarettes" (77–78). The threat signaled by the presence of lit cigarettes is reined in—fire does not occur at this point in the novel—but its mention is enough to suggest a latent threat in the volatile gases released along with the oil. When the danger implicit in this passage does erupt it is fitting that it comes via the cataclysmic scene I discussed earlier; the fact that Bunny's first oil well explodes so spectacularly suggests a liaison between risk and verticality in order to establish the entire enterprise of oil drilling as one founded upon instability.

The passage above takes on an unsavory flavor but one concerned with literal griminess; "gas," "oil," and "mud" repeatedly return the prose to a basic and unembellished idea of dirt. In fact, the extraction process is rendered in poetic terms, so that the narrative ultimately revels in the dirty work of drilling. The half-rhyme of "dump it into the sump-hole" and the embellished phrase "a lovely dripping black" are entirely at odds with the more restrained, scientific terms "pressure," "a mile deep," and "a bucket fifty feet long." By playing with these formal and exact terms and then occasionally breaking into more excited language, the passage has a discordant timbre. In the passage, this uneven tone suggests something about the underlying contentions at work in Sinclair's representation of verticality; oil drilling produces a sense of aesthetic value in the "lovely dripping black," but is also associated with banal industrial work. The tension between bland language and an animated poetic form joins the implication that oil drilling has an underlying potential for mortal danger; here the suggestion of a conflict between spectacle and danger is supplemented by that between aesthetic pleasure and industrial ordinariness.

Cultural texts such as Edward Burtynsky's photographic collection *Oil* (2009), which documents oil extraction among a haunting and ruinous landscape, compound this clash between what is both aesthetically rousing and visually threatening. Burtynsky's *Oil* is relentlessly preoccupied with a compulsive national desire for oil. Its visual lexicon, like the tenor of the *Deepwater* accounts with which I opened this chapter and Sinclair's depiction of roiling black smoke covering the burning oil derrick, rests upon the evocation of disquiet. The image titled "Burning Tire Pile #1" (1999), for example, captures the immediate spectacle of smoking mountains of waste to connote the catastrophic social and human consequences of these same scenes. Sinclair's narrative, sharing the same essential dynamic, also enters into a dialogue with the disaster spectacles I discussed in chapter 1. Popular in the years immediately prior to the publication of *Oil!*, the disaster spectacles included staged fires that engulfed tall buildings and consciously eschewed "the convention of the 'removed' audience" by demanding audience participation without placing them in any real danger.[24] The formal difference between the performative disaster spectacle and *Oil!*'s textual inferno does not preclude an understanding of how both participate in a wider contemporary context that culminated in the film industry's establishment of the disaster film "both as a spectacle and as a narrative genre," and which Bill Brown argues effectively made disaster "the privileged mode for effecting the recreational sublime."[25]

Brown concludes that, by "effacing the narrative frame," the disaster spectacle could "evacuate the ethics of narrative" and "purify the sensation of cognitive complication" for its audience. In this way, the spectacles provided a "visceral reaction dislodged from moraliz(ing) technology"; "in the words of modernism," "Coney Island realism had derealized the content of disaster: subject matter as such had become the mere motivation of sensational effect."[26] Brown uncovers the inherently problematic voyeurism involved in consuming representations of calamity, but, while the disaster spectacles may have allowed their audience to observe destruction without feeling complicit in a real catastrophe, the performances did not truly eliminate an ethical framework from representation. *Oil!*'s textual depiction of a disaster spectacle is notable for not attempting to remove the ethics of complicity that always haunt the portrayal of destruction. Rather, the novel makes verticality an expression

---

24. Dennett and Warnke, "Disaster Spectacles," 104.
25. Brown, *Material Unconscious*, 118.
26. Ibid., 116–18.

of the very conflicts involved in both the depiction of, and the motivation for depicting, catastrophe. Throughout *Oil!*, vertical imagery cultivates a sense of precariousness that rests upon the activation of various conflicts in tone (excited and anxious), voice (satirical and affectionate), and image (catastrophe and banal physical labor). The crossover between the inherent dissonance of the image of a burning oil derrick in Sinclair's novel and the problematic ethics suggested by the Coney Island disaster spectacle uncovers how the ambiguity of moments in which apparently incompatible ideas coexist allows *Oil!* to foster an underlying sense of uncertainty in the face of industrial advance.

THE OIL DERRICK AND THE SUBLIME EXPERIENCE

In her description of viewing a loop comprising aerial views of "the petrochemical industries on Houston's Shipping Channel," Stephanie LeMenager names a "petrochemicalscape" that "becomes more lunar," with "rows upon rows of brilliant white oil drums." As she watches, the "sound track grows nervy, building toward buzzing, cacophonous dissonance." The video causes LeMenager to state:

> An evocation of the petrol sublime, these moving images are less reassuring than the Romantic mountain or waterfall because their movement is both humanly motivated and illegible, completely undercutting the confidence that the sublime was supposed to produce—that we humans actually could glimpse the infinite, God's scale. My friend and I identified occasional trucks and trains crosscutting the industrial maze with glee, seeing in them a vague hint of plot.[27]

In its key statement—that this contemporary vision of oil procurement fails to produce the "confidence that the sublime was supposed to produce"—LeMenager's critique reminds that the power of the sublime is often attached to, or even dependent on, a degree of unknowability. Perhaps we have become blind to the visual impact of the oil refinery or derrick. As Andrew Pendakis writes, "Our conception of oil is usually oriented by [the] wide-angle image of the silently running oil refinery or platform"; oil becomes, in this manner, "dangerously literalized, wrongly

---

27. LeMenager, *Living Oil*, 139.

conceived as simply coextensive with a highly simplified figure of its own productive apparatus."[28]

In contrast to these contemporary scenes, *Oil!*'s vertical sublime operates within the framework of the skyscraper's relative newness and, in this newness, generates a sense of unknowability lacking from the now-familiar scene of oil procurement. The following examines how the conflicts that were the emphasis of this chapter's previous section congeal in the image of the oil derrick, which is placed in distinction to the more familiar vertical structure of the tall city building. Sinclair's novel is already unique for its use of the derrick and oil refinery as industrial images among narratives that typically use the factory or railroad to interrogate industrial progress.[29] Moreover, the derrick is posed in *Oil!* as a new way of interpreting the American technology that allowed tall buildings to rise into the sky. Oil derricks, as well as the oil well shaft and the geysers erupting from them, suggest capitalist endeavor as an explosive and intrusive force. They are also a crude counterpart to the skyscraper; both resonate with technological progress, but the derrick's architecture was determined more by practical concerns than the aesthetic dictates of the new skyscraper.

The derrick is an altogether more conflicting symbol of technological advance. It suggests industrial progress and yet most necessarily exists within an unpopulated landscape, making it appear as a cipher of plenitude in the wilderness. In a photograph taken around the time that Sinclair was writing *Oil!*, oil derricks crowd into a barren landscape that lacks any sign of human habitation.[30] The derricks recede into the distance in a composition that is at first glance banal; on closer inspection, a multiplying vista of vertical structures—not dissimilar to the *mise en abyme* of skyscrapers greeting Nemo—recedes into the distance. Like the replicating skyscrapers in *Little Nemo,* the repetitive verticality of the derricks in this photograph elicits a sense of uncanniness. Moreover, the photograph's visual arrangement works in a way that makes the oil derricks dwarf the distant mountains; the derricks produce an overwhelming verticality exceeding that of the skyscraper pictured in an already densely

---

28. Pendakis, in Biemann and Pendakis, "This Is Not a Pipeline," 8.

29. Both Marx, *Machine in the Garden,* and Nye, *American Technological Sublime,* prioritize these structures and suggest them as particularly significant images of American industry.

30. The photograph I refer to here is titled "An Oil Field" and was created between 1900 and 1920. It is held by the Library of Congress Prints and Photographs Division and is available to view online (reproduction number LC-DIG-det-4a25710).

vertical urban world. The subtle replicative visuality of these derricks has a lingering effect that amplifies the force of the photograph, which rests upon an underlying contrast between the visual presence of the derrick and the deserted space, and between the referential suggestions of these two visual symbols—namely, industrial progress and wilderness. *Oil!* uses the derrick as an appropriation of the skyscraper's form to explore the dichotomies that are visually summoned by such contemporary photographs and channels these conflicts into an appraisal of the inherently problematic nature of industrial progress. Throughout *Oil!*, the drilling that takes place at the derrick is rendered in a discordant tenor, suggesting the unsavory allure of both monetary profit and the excitement afforded by the sight of oil spurting from the ground.

Nonetheless, the novel does not condemn oil drilling itself so much as what it stands for; in fact, oil drilling itself is valorized as both an exclusively masculine pursuit and also an activity that energetically fosters this same exclusivity. The appeal of oil drilling is cemented by *Oil!*'s creation of a kind of surrogate familial structure that suggests underlying sympathy for the masculine domain of industrial labor. The kernel of this male utopia exists in J. Arnold Ross and Bunny, a family unit in which Bunny's mother is absent, his grandmother a figure of gentle ridicule, and his sister characterized as a superficial young woman who is to be disregarded. From the seed of father and son, Ross Consolidated multiplies within a "fine new Spanish 'ranch-house'" in which "the executives and directors and geologists and engineers of Ross Consolidated" form an entirely male "big happy family" (255). Sinclair's representation of oil drilling bears the kind of veneration tangible in contemporary photographs of oil workers in which groups of men, sometimes workers, sometimes businessmen, gather around the wooden frame of a derrick.[31] The composition of these photographs, as well as the men depicted within their frames, articulate a sense of pride; the figures that populate the image are carefully framed and stand straight-backed, or with their hands pocketed in affected nonchalance. A similar eulogization of the hard labor associated with industry emerges in images of skyscrapers from the same period, the most famous of which are the photographs by Lewis W. Hine.[32]

Hine's photographs have a large scale that creates a heightened impact when using vertical perspective, but the similarities between the

---

31. See, for example, the image of an oil derrick c. 1880s, reproduced by Morritt, *Rivers of Oil*, 103.

32. See also Hine's "Men at Work" series in Hine, *America and Lewis Hine*.

skyscraper and oil derrick photographs indicate a wider trend in which verticality stands in for a particularly male, and often dangerous, industrial endeavor. The male world of Ross Consolidated at the site of the oil derrick reveals how Sinclair's novel summons the same relationship between verticality and masculine labor. Furthermore, *Oil!* suggests that there is something worthwhile and respectable in menial work; throughout, the narrative conveys palpable reverence for the workers at the derrick, and Bunny eventually concludes that the moral integrity of a worker has greater allure than the power and wealth waiting for him should he fulfill his father's legacy by becoming an oil man. The celebratory sketches of the proletariat in *Oil!* also determine the narrative's trajectory, which culminates in Bunny's final movement away from the oil business in order to establish his own workers' school, a reflection of Sinclair's own intentions to propose a Socialist future for American industry. While the novel has an underlying satirical edge directed toward the greed of both the big "oil men" and the oil companies, it seems that Sinclair's political agenda is so entrenched that he cannot entirely condemn oil drilling, since such a move would consign those who conduct its labor to a similar fate.

Although *Oil!* often veers between overt criticism for the oil industry as a whole and an underlying sympathy for the microcosmic world of drilling, this unevenness does not undermine what is a consistent vocalization of the simultaneous allure and danger of American industry. If the oil derrick is used as the manifestation of a valorized masculine world as well as a structure that recalls the skyscraper and conveys disquiet about industry, the novel ties vertical oil structures to ideas of the sublime in order to worry further at the conflicts between pleasure and anxiety. Placing a similarly high premium on industrialization as Leo Marx and Alan Trachtenberg, David E. Nye suggests that the distinctly American form of the sublime was a direct consequence of the "throes of rapid industrialization" and "geographic expansion" that are so central to *Oil!*.[33] By making the oil refinery the site of Bunny's inauguration into a Socialist way of thinking, *Oil!* suggests that the verticality Bunny sees causes a revelation akin to that produced by the sublime scene. The oil refinery is the specific place where Paul Watkins reveals to Bunny how "the modern industrial world with its manifold injustices" can be transformed via Socialism into something capable of beauty and heroism. With the vertical building before him, Bunny sees himself as "a man lost in a tangled forest," who is "taken up in a balloon, and shown the way out of the tangle"; Paul's

---

33. Nye, *American Technological Sublime*, 1.

ideas make everything "simple, plain as a map." Bunny has the revelation that "the workers were to take over the industries, and run them for themselves, instead of for the masters" (275–76). The narrative rests upon a vertical metaphor in order to communicate this revelation. The figurative ascent suggested by Bunny's political enlightenment segues into a literal reflection upon the "new refinery," which is a "great building," a "wonderful work of art" that rises before them. This "great" structure condenses the metaphor of Bunny's balloon flight into a concrete vertical fact (276). That the oil refinery stimulates Bunny's political reformation suggests a relationship between the visibly vertical and the transformative experience central to the sublime.

It seems that Sinclair was considering the verticality of drilling and mining as the basis for a fictional consideration of the oil industry from 1908, when he went "wandering" in the Southwest. Here the author witnessed firsthand the "towering peaks" of Bishop, California, and the "greatest gold mine" in America, a "vein" that went "all the way up the mountainside," with a mining camp founded on a "basis of comradeship, with high wages and plenty of socialist propaganda." Sinclair soon realized that most of the product was "low-grade" and only profitable if it could be mined on a large scale, which in turn revealed how the "complications" of quartz mining forced the owner to turn his mine over to "the big capitalist groups that sought to buy it and freeze out the old stockholders."[34] I propose that this encounter in 1908 prompted the beginnings of *Oil!*, allowing Sinclair to see how the underlying moral problems of oil drilling as part of a wider capitalist market could be expressed through vertical imagery. If there was an inherent spectacularity in the scene Sinclair witnessed of a "towering mountain" and a "vein" running up its side, there is an even more explicit emphasis on the feelings generated by visually appreciating the vertical structures and events of oil procurement in *Oil!*.

Bunny's description of the oil refinery is both full of awe and whimsical; the building is a "wonderful work of art" and simultaneously a structure "made entirely of enormous baking-pans" that causes him to acclaim, "The angels were making caramels for the whole world, dainties with a new, patented flavor." Now that Bunny views Socialism as morally expedient, it seems that drilling has lost the basic dirtiness it had in the initial description of an oil geyser. It is twilight, and the scene confronting

---

34. Sinclair, *Autobiography*, 147–68.

Bunny and Paul is apparently magical; "the white steam" rising from the buildings has "a faint violet tinge" as it blends with the sky. Finally, "white and yellow and red" electric lights blink on, so that the place looks "like a section of Coney Island" (276). Coney Island "temporarily overturned and rewrote the social order" by upending values of "thrift, sobriety, restraint, order and work" and instead "exploiting technology for pleasure." In these terms, Sinclair's comparison between an oil factory and amusement park is apt; it was in the same period that Henry Ford's River Rouge Plant became a "popular tourist site" due to its display of the assembly line.[35] Coney Island also has particular cachet as a site of overwhelming verticality; it offered Sir Hiram Maxim's Flying Machine, which spun elliptical airships at tremendous speeds in an exhilarating sensory experience. As in *Little Nemo*'s visual allusions, the reference to Coney Island in *Oil!* works to join vertical spectacle with physical experiences removed from the parameters of everyday life.

Earlier in this chapter, I suggested that the textual representation of the oil fire that Bunny witnesses recalls the disaster spectacles of American amusement parks in that it also allows catastrophe to be consumed within the parameters of fictional narrative. In its fictionalized oil well fire, *Oil!* detaches the disaster represented from a historical moment in which real lives are lost in ways that other texts, such as the post-9/11 narratives I discuss in chapter 5, are unable to do. But even in its fictional disaster, *Oil!* enters into a fraught contract in which the readers must negotiate their own pleasure at the imagination of great destruction. Sinclair's novel does not attempt to excise the ethics of either depicting or consuming disaster that inevitably haunt the portrayal of destruction; rather, it makes verticality into an expression of the conflict involved in the description of catastrophe. Further, the novel uses verticality as a clash of those opposing forces and tensions I have discussed to imply that the ethics of watching (and enjoying) disaster are similar to those involved in consuming (and taking pleasure in) the products of capitalist endeavor.

Bill Brown suggests that in some ways the disaster spectacle tried to fix contingency into a predictable sequence for its audience, just as mining fixed nature into a sequence of locations to procure economic resources. He argues that the "serial reproduction of disastrous destruction" at the fin de siècle marks the moment "when the amusement industry routinizes the aleatory," comparable to the way that the mining industry had

---

35. Nye, *Electrifying America*, 127–29; idem, "Accelerating American Experience," 10.

"mechanized nature."[36] The sublime moment precipitated by the vertical in *Oil!* is both an echo of, and works in radical distinction to, the temporality in Brown's analysis of the disaster spectacle. Looking up at the oil refinery summons "swift flashes of revelation" for Bunny (276); the vertical structure produces punctuating moments of understanding. These revelatory moments enable Bunny's transformation and, in their temporal punctuation, accord with the key register of the sublime: what Nye refers to as the "stepping out of historical time into the eternal now."[37] In this sense, Bunny's revelation in *Oil!* does not standardize the aleatory so much as affirm the concept of the spectacle as without the normal structure of time. Writing about the emergence of spectacle as an "attempt to deal with the temporal instability of the image," Mary Ann Doane notes that spectacle is "not the taming of the contingent, but its denial."[38] By the same measure, however, the idea that the spectacle and the sublime either step outside time or reject contingency relies upon a conception of temporality; even as these models try to deny time, they inevitably invoke its centrality.[39] In this sense, the scene of the oil refinery accords with the novel's ongoing contradictory impulses; the problematic structure of time inherent in the model of the sublime moment and the disaster spectacle become another way in which the novel fixes various paradoxes to vertical imagery.

Ultimately, Sinclair's depiction of the sublime retains the central disquiet that accompanies the overwhelming experience of viewing vertical events. The sensory delight attached to the scene at the oil refinery is tempered by the more complex pleasures at work that inscribe verticality as the source of both aesthetic consumption and anxiety. The passage's metaphors of domesticity and theme parks—the sweet foods and blinking lights—transform the more violent verticality of other sections of the novel into sensual enjoyment while retaining the sense that something less appetizing lies beneath. The oil factory, promising visual pleasure and harboring danger, is a site of the technological sublime; the vertical imagery of oil drilling rests upon the natural sublime in Sinclair's description of an oil geyser as "a roaring and rushing, as Niagara." This analogy suggests an eagerness to marshal the associations of the natural sublime in order to express the overwhelming scale of the oil geyser, but Sinclair's simile both places the event as similar to, and a direct inverse of, Niagara

---

36. Brown, *Material Unconscious*, 118.
37. Nye, *American Technological Sublime*, 25.
38. Doane, *Emergence of Cinematic Time*, 170.
39. Mickalites argues similarly in "Manhattan Transfer," 60.

Falls: the geyser is made of oil, not water, and shoots "two hundred feet, two hundred and fifty" upward into the sky rather than rushing down into a cavernous lake (25). By suggesting the oil geyser as an unfamiliar echo of a familiar example of the natural sublime, the novel effects a sense of oil drilling as a source of the technological sublime and an activity that is at odds with the natural world. Moreover, the moment in which Bunny observes the oil refinery with Paul makes the novel stand apart from contemporaneous texts such as those by William Dean Howells and Robert Herrick, which also demanded social reform, but did so by either underplaying—or denying altogether—the sublimity of the industrial scene.[40]

Oil!'s invocation of the sublime attaches both supernatural experience and social critique to the verticality of oil geysers and factory buildings. The scene of the oil refinery suggests the mutability of vertical metaphor; the vertical scene is a harmless, almost fairytale-like site of moral transformation for Bunny, while it also reveals a critique of the oil industry as a whole. Despite the superficial pleasure suggested by the buildings as forms of confectionary, they give off "sickish sweet odors that spread over the hills for miles." These smells are so alarming that they even frighten other life (namely, quails) away (276). The pervasiveness of those "sickish sweet odors" alludes to another "sickish sweet" smell—that of blood—opening the door onto a wider parallel between violence and verticality. The forbidding undercurrent suggests that the refinery, while superficially somehow pleasing, is marred by the cost of human labor, not dissimilarly so to the depiction of the industrial meat packing district in Sinclair's *The Jungle* (1906).

In this earlier novel, Sinclair describes "two rows of brick houses, and between them a vista: half a dozen chimneys, tall as the tallest of buildings, touching the very sky." These monuments are touched by "half a dozen columns of smoke, thick, oily, and black as night." Sinclair's description culminates in the suggestion that the smoke "might have come from the center of the world," "where the fires of the ages still smolder"; it comes "as if self-impelled, driving all before it, a perpetual explosion."[41] The portrait of the oil refinery in *Oil!* extends the imagery in *The Jungle,* also marked by an almost apocalyptic depiction of fire and dirty smoke. If *Oil!* is tempered by Bunny's light-hearted description of the oil refinery, both texts summon the darker elements of industry and

---

40. Nye, *American Technological Sublime,* 123ff.
41. Sinclair, *Jungle,* 28.

make verticality an articulation of the ongoing conflict between the joys afforded by wealth and the costs of indulging in this pleasure.

DRILLING INTO SUBTERRANEAN/SUBCUTANEOUS SPACE

The lingering traces of violence and immorality in *Oil!* exemplify the novel's consistent return to the effects of capitalism and the way in which it tends to evaluate these consequences in terms of human cost. This tendency is evident in the novel's climax, which depicts Bunny rejecting the oil industry precisely because of its detrimental effect upon the menial laborer. It shadows the entire novel, throughout which the specter of physical danger looms. It also informs specific moments that suggest men are transformed into automatons by a career in the oil industry, such as the image of Ross as a "complicated machine" with "no romance in him," a man whose emotions are like "a furnace" that takes "a long time to heat up" (7; 14). This last example reveals a network of imagery that creates a crossover between the earth and the human body. The final part of this chapter examines how the image of Ross's body is interchangeable with the machinery he operates, how this imagery is paralleled by the inverse image of oil drilling as reminiscent of penetrating the human body, and how this latter image is facilitated by a liaison between oil and bodily fluids.

When *Oil!* ties verticality to masculinity, it implies that maleness is inextricable from violence, an idea *There Will Be Blood* graphically depicts by using a vertical inferno as a symbol of volatile masculine behavior. What is more explicit in *Oil!* is the relationship between the human body itself and verticality. The vertical reach of the derrick has an inevitable phallic resonance; the erupting geyser of oil and the thrust of the drill into the earth cannot escape their suggestiveness. The first time oil spurts into the sky it is a particularly sexual image, made even more so by the jubilant tone of the prose:

> There she came! There was a cheer from all hands, and the spectators went flying to avoid the oily spray blown by the wind. They let her shoot for a while, until the water had been ejected; higher and higher, way up over the derrick—she made a lovely noise, hissing and splashing, bouncing up and down! (78)

The oil spurting forth is rendered androgynous: the geyser is a "she," but the image of an "oily spray" shooting from the earth is unmistakably reminiscent of male ejaculation, compounded by the verbal echo of "ejaculate" in "ejected." The feminization of the geyser, combined with the parallel between oil and semen, makes the oil drilling depicted seem to allude to heterosexual sex. "There she came" is directly reminiscent of the language of Herman Melville's *Moby-Dick* (1851), also an oil narrative, but one depicting the earlier method of procuring oil from whale blubber. *Moby Dick* climaxes with the proclamation "There she blows!—there she blows! A hump like a snow-hill" as the men on deck watch the sight of the titular whale showing "his high sparkling hump, and regularly jetting his silent spout into the air."[42] Sinclair mirrors Melville's female pronoun and the jubilance of his prose, referring back to both an originary rapture in the face of natural phenomena and an antecedent oil narrative; *Oil!* both heightens the joyous sexuality attached to the oil geyser and places the scene in dialogue with a longer tradition of representing oil procurement.

The sense that oil drilling is a recollection of sexual acts is compounded by the general preoccupation with sexuality in Sinclair's work. His early writing has been described as "consciously" repressing "any kind of eroticism"; in these terms, the battle scenes and sieges in the Civil War epic *Manassas* (1904) are read as a "shift" of the author's "sexual energy" to "other fields."[43] Sinclair's autobiography attests to as much via moments of autopsychoanalysis in which the author declares that he "learned to work fourteen hours a day" in order to keep "the craving for woman" at bay.[44] In *Oil!* Bunny Ross's sexual curiosity allowed Sinclair to explore issues that, for long periods of his life, he was not able to confront and seems to affirm that the author's creativity was an extensive process of sexual sublimation. Significantly, the overt sexual tones of the passage above depicting the geyser of oil contrast starkly with the veiled euphemisms of Bunny's first encounter with sex:

> She put her lips on his, in a long kiss that made him dizzy. He murmured faintly that something might happen, she might get into trouble; but she told him not to worry about that, she knew about those things and taken [sic] the needed precautions. (200)

The prose continues with a vertical metaphor in which sexual intimacy is "walking on a slippery ledge, over a dark abyss where pleasure and pain

---

42. Melville, *Moby-Dick*, 538.
43. Hornung, "Literary Conventions," 26–27.
44. Sinclair, *Autobiography*, 46–47.

were so mingled you could hardly tell them apart" (201). Like the image of the well spurting oil, physical intimacy is afforded the language of verticality, although more obliquely, with suggestions of being made "dizzy" and the images of crossing a "slippery ledge" with an "abyss" below. Here the tone is conspicuously subdued, scattered with opaque references to pregnancy as "something" and "that," and marked by conditional phrasing such as "might." The underlying relationship between sexuality and creativity in Sinclair's work and the way that his prose refracts ideas of sex into other narrative gestures such as oil drilling perhaps explains why *Oil!* has such an enigmatic tone.

While Sinclair's tug-of-war between extremes of chastity and sexual desire in his own life means that verticality is a conspicuous sexual metaphor in the novel, these observations should not distract from the more significant fact that the novel has a deeply rooted concern with the relationship between verticality and violence. In *Oil!*, the fluid that forms the ultimate goal of drilling is made inextricable from the vertical process of obtaining it in a central narrative emphasis on oil as a buried thing. While the passage above detailing the geyser's oily spray suggests oil as semen, elsewhere the imagery suggests blood. Oil is depicted both deep within the earth and shooting from it in a spectacular geyser, allowing it to function as a metaphor for two interrelated bodily fluids:

> The inside of the earth seemed to burst out through that hole; a roaring and rushing, as Niagara, and a black column shot up into the air, two hundred feet, two hundred and fifty—no one could say for sure—and came thundering down to earth as a mass of thick, black, slimy, slippery fluid. (25)

The passage continues with a description of the oil "carried by the wind, a curtain of black mist" that "sprayed the Culver homestead, turning it black." It concludes with the exclamatory assertion of a "million-dollar flood of 'black gold'!" (25). Oil is "the inside of the earth" and, like the inside of the body, is "thick, black, slimy, slippery." The developing parallel in the passage between oil and blood means that the passage's final reference to "black gold" implies that the spilling of blood is the true cost of heavy industry.

The early suggestions of danger indicated by the potential ignition of the oil geyser, and the later discomfort associated with the ominous façade of the enormous oil refinery buildings, culminate in the suggestion that the ultimate cost of oil drilling is human life. The narrative correlates the desire to plumb ever deeper depths into violent consequence; at Bunny's well, the decision to drill "deeper" almost immediately meets with an "accident" as one of the workers reports that, "There's a man fell in the hole" (151–52). Ross quickly deduces that, since there is no sound emitted from the well, the man must have "drowned in the mud." He then exclaims, "My God! My God!" and "It makes me want to quit this business!" but his grief is tempered by its repetitiveness. Bunny has "heard that cry a thousand times before." For Bunny, the accident is translated into the effect it has had on his enjoyment of drilling; he becomes "sort of sick inside"; and the narrative reflects, "Gee, what a shame—in his well, of all places, his first one," and "It was all spoiled for him; he wouldn't be able to enjoy his oil if he got it!" (152–53). Soon past his apparent distress, Ross efficiently instructs the workers to send down a three-pronged hook, meeting the foreman's concerns that this might "tear" the man with the declaration, "You've got it to do. It ain't as if he might have any life in him. Bend the prongs so they fit the hole, and force them past the body. Go ahead and get it over with" (153). The man is now merely a "body," nominally marked as a kind of sacrifice to the wider operation that is being conducted. It is as if, in order to successfully extract oil from the earth, the workers must give up a human life, further blurring the distinction between oil product and its labor.

The moment is prescient of LeMenager's declaration that, through the force of petroleum products, we now *live* oil; our existence is mediated by objects derived from oil; and "there can be no 'liveness' without mediation." Oil also "challenges liveness from another ontological perspective," because it is a substance "that was, once, live matter and that acts with a force suggestive of a form of life."[45] LeMenager continues to argue that "energy" functions as "a metaphor that obscures our laboring bodies, offloading work as a grounding concept of our species onto other entities, such as water, wood, coal, and oil." In Sinclair's novel, "oil itself returns, with almost every representation of its discovery, as an excessively embodied figure, the viscous medium of unregulated play"—oil in Sinclair's novel is, in fact, "apparently more alive than its human witnesses."[46] The man's descent into the bowels of the earth in *Oil!* crucially indicates how verticality frames the novel's wider concerns: here verticality is the movement that attenuates the equation between human blood and oil. Although the man's body is recovered from the oil well, his death takes

---

45. LeMenager, *Living Oil*, 7; 6.
46. Ibid., 191; 93.

place at the bottom of the shaft so that it seems he has been swallowed up by the earth. The scene extends the blur between the bodies of workers and the commercial product they supply; the implication is that human flesh will now mix with, and contribute to, the oil extracted by Ross.

There Will Be Blood makes this conflation of oil and blood visual. The film's title announces that blood will be spilled at some point during its course, but the narrative emphatically and repetitively depicts the eruption of oil out of the ground, and the suggestion that blood will be shed from a human body is initially displaced by the image of oil bleeding from the earth. While the film is framed by two graphic deaths—that of a man falling down the oil shaft at the film's beginning and the murder of Eli Sunday by Plainview at its close—the first of these moments crucially aligns the color and viscosity of crude oil with that of blood. As Daniel Plainview stands at the top of the oil well with other workers, a piece of machinery accidentally drops down the shaft and kills a man, who is shown at the bottom of the well covered in oil and mud while the next shots show Plainview's face. Because the man's body is covered by oil, any evidence of blood is obscured, allowing the film to syntactically relate the man with Plainview, who is also covered in dark fluid. Making blood and oil interchangeable suggests that a symbolic blood must be spilled in order for the commercial product of petroleum to be obtained. By *There Will Be Blood*'s closing scenes—in which red blood is finally shown, unobscured by oil and made stark against Eli Sunday's pale skin—it is as if the film's titular promise has already long since been fulfilled.

In *Oil!*, bloodshed occurs in a mainly allusive form in the scene of the worker falling down the oil well, but his death crucially operates as a symbol for a larger community of those exploited or neglected people whose blood is spilled at the hands of American industry. This becomes apparent during the second instance of a body falling down an oil well; the man's death causes Ross to recall a less grievous, although possibly more disturbing, accident at "the first well he had drilled," where he witnessed a baby fall down a shaft. Incredibly, the baby survives:

> They fished for it with the big hook on the end of a rope, and got the hook under the baby's body and lifted it gently a few feet, but then the body got wedged somehow, and they were helpless. The child had hung there, not screaming, just making a low, moaning sound all the time, "U-u-u" like that, never stopping; they could hear it plainly. They started twenty feet from the well and dug a shaft, big enough for two men to work in, breaking the ground with crow bars, scraping it into buckets

with big hoes, and the men on top hauling the buckets out with ropes. When they got below the baby, they ran in sideways, and got the baby out all right. The hook had sunk into the flesh of the thigh, but without breaking the skin; the bruise had healed, and in a few days the child was all right. (154)

In narrative terms, the horror of the man's death seems to have been accelerated by the image of a baby falling down the well; the even more vulnerable body heightens the sense of what the man's death means. The fact that the baby's survival is so incredible, and strays so thoroughly from the novel's realist mode, suggests its importance within the plot. Although, or perhaps because, the incident takes place beyond the borders of the narrative's chronology—it interrupts, from Ross's own past, into a narrative largely tied to Bunny Ross's experience—it has great impact upon the narrative. The ultimate rescue of the baby is less significant than the horror of the depiction of its ordeal. And the baby's survival does not undermine the passage's sense of inevitability, manifest in the repetitive return to the words "baby" and "body," which occur in various forms at least six times in the prose and are conjoined by violent alliterative partners such as "big hook," "breaking," and "bruise." Crucially, both the baby's and the man's falls stand in for two implied wider costs; the first relating to the earth, which is being plundered for its natural resources, the second, more emphasized, relating to the community of bodies killed in the First World War.

The extraction of the man's body from the well prompts the narrative, in free indirect discourse approximating Bunny's voice, to declare:

> Your thanksgiving dinner was spoiled, because one poor laborer had slid down into a well which you happened to own; but dozens and perhaps hundreds of men had been hurt in other wells all over the country, and that didn't trouble you a bit. For that matter, think of all the men who were dying over there in Europe! All the way from Flanders to Switzerland the armies were hiding in trenches, bombarding each other day and night, and thousands were being mangled just as horribly as by a grab in the bottom of a well; but you hadn't intended to let that spoil your Thanksgiving dinner, not a bit! (155)

With this passage, the meaning of the man's body as a member of a larger community is revealed. The "thousands" of soldiers "being mangled just as horribly" make the trenches parallel the oil well and suggests two things: that oil and war are inextricable and that the vertical descent of human bodies into the earth is the primary metaphor for the novel's political concerns. Throughout, *Oil!* emphasizes oil as crucial to military victory; Bunny Ross realizes that his father possesses the product that will win the First World War as the narrative declares unequivocally that "what was going to win the war was bullets and shells" but in order "to get them to the battle-field you had to have transportation"; the oil that Ross brings "up out of the ground" is "driving big trucks" that are "carrying munitions up to the front"; it is "moving the biggest and fastest cargo-ships"; it is "lubricating the machinery in the factories"; and "more and more" is being demanded (205).

The phrasing of this last clause reiterates the early association between oil and acceleration I noted in the depiction of the car journey. "More and more was being called for" has a neat repetition of "more" as well as a near repetition in the rhyme of "more" with "for"; the specifically poetic tone of the depiction captures the repetitiveness, and therein the compulsiveness, of a need for oil. Ross even declares, "The world has got to have oil," a statement that both resounds definitively and simultaneously suggests oil is addictive (292). "Got to have" suggests compulsion both in its syntax, which is succinct and direct, and also its semantic message, which uses "got" and "have" to imply the sheer desperation of needing oil. A 1958 poem also entitled "Oil," written by American poet Gary Snyder, neatly distills the suggestion that oil is like a drug through the description of a ship's load as that which is required by a host of addicted nations, namely: "long injections of pure oil."[47] The poem shares the modernist experimentalism of William Carlos Williams and is formally and temporally distant from Sinclair's novel. But reading the texts together ratifies oil as a recurrent symbol in twentieth-century American writing and foreshadows the way in which verticality is attached to technological and industrial advance by Snyder's contemporary, Allen Ginsberg, whose work I discuss in chapter 4.

Snyder depicts a congregation of laboring, masculine bodies to suggest that these various human bodies constitute a community essential to the poem's topic, which is only revealed in the final line as "pure oil." In *Oil!*, the creation of a male community surrounding oil drilling is also tied to the extraction of the product. Snyder's poem realizes two ideas that Sinclair often seems to suggest: both the image of oil spurting forth from the earth as an almost sexual ejaculation and the suggestion of oil as an

---

47. Snyder, *Range of Poems*, 120–21.

addictive drug. Its final stanza briefly summarizes an overarching sentiment of Sinclair's through the depiction of the world as addicted and the suggestive double allusion to "long injections of pure oil," where "long injections" might equally refer to injections into the earth or hypodermic injections into the human body. The poem betrays a pervasive analogy in which the products of capitalism are imagined as addictive, while echoing the correlation between industry and images of vertical intrusion. In *Oil!* the drill's penetration of the earth and the image of bodies falling into a deep well rest upon the same correlation, notably shedding the ambiguity of the novel's moments of simultaneous pleasure and anxiety, and leveling a resonant warning against the potential consequences of industrial progress.

This chapter has examined how *Oil!* engages with the pleasurable aspects of drilling, extolling the visual impact and sensory exhilaration of seeing a geyser erupt from the earth and reveling in the physical experiences afforded by the oil industry, such as car travel, while simultaneously invoking a conflicting sense of danger or anxiety. While *There Will Be Blood* can be read as a text about exhaustion—both the dwindling of frontier expansion and the "tiredness" surrounding "twentieth-century stories about modernity"[48]—*Oil!* similarly signals a sense of fatigue surrounding the depletion of horizontal space and, in its stead, offers up the spectacularity of vertical imagery. In this sense, it foreshadows, but is very different in tenor from, the spatial concerns of Native American literature discussed in the next chapter. *Oil!* condenses the various conflicts articulated by its plot, thematic emphases, and its tenor through verticality. As imagery that is mobilized in order to communicate the novel's political agenda as well as fulfill the dramatic tone of both the prose and the plot, verticality suggests the allure of oil and the danger associated with its unchecked use; while *Oil!* suggests that peril and violence lie beneath the surface of American industry, it simultaneously embraces the sense of awe that accompanies extreme vertical events and structures such as derricks and oil geysers.

Sinclair's novel marshals vertical imagery in order to actively summon, although not dispel, anxieties and equivocations about the oil industry and the capitalist project to which the same industry metonymically refers. *Oil!*'s clash of opposing narrative registers and tones—critiquing its protagonists one moment only to affectionately extol them the next—uncovers a central textual ambivalence, and Sinclair's verticality translates the apprehension captured by the dream-comics into an expression

---

48. LeMenager, *Living Oil*, 100.

of the inherent difficulty of appraising broad cultural movements and ostensible progress. Ultimately *Oil!* is resolutely unwilling, and seemingly unable, to render the technological and industrial advances it depicts in wholly positive or negative terms. Precisely because verticality in *Oil!* is an elaborate and ambivalent construction, the novel articulates a repetitive struggle with the desirability of emerging American capitalism. And because *Oil!* stands in distinction to other texts that offer verticality as a fixed symbol of modernity (whether positive or negative), it uncovers verticality as a way of mediating a contemporary uncertainty about the future of capitalism for which the oil industry, with its short-term and highly visible destructive aftermath, stands. The crossover between human blood and oil as part of the novel's central idea that drilling is inextricable from the First World War foreshadows the comparable, but slightly different relationship, between oil and the late-twentieth century wars in the Middle East. Sinclair's writing marks an emerging concern with an age of physical verticality and its relationship to political life, which is the focus of the following chapters.

The environmentalist Paul Relis described witnessing the 1969 Santa Barbara oil spill from a small plane in these terms:

> It was a transformative experience, because I'm literally looking down into the upwelling of oil, oil surging in the ocean. I was mesmerized: I had never seen anything like it. And it hit me that this is a world-changing event. It was not going to be the same after this, whatever "after" is.[49]

Relis uses terms strikingly similar to those of firsthand descriptions that I take as the primary subject of chapters 4 and 5: descriptions by the air crew on board the plane that dropped atomic bombs in Nagasaki and Hiroshima in 1945, Philippe Petit's description of standing at the top of the World Trade Center in 1974, and eyewitness accounts of the terrorist attacks on 11 September 2001. The sense that time stands still in the witnessing of a vertical event is an iterative trope throughout the texts discussed in this book. In *Oil!,* the scene at the oil refinery recalls the inherent contradictions of the sublime moment as in and outside of temporal structure; Bunny waits, not so much for the sky to fall, but with his anticipative gaze marked by a sense of apprehension and delight concerning the vertical spectacle before his eyes.

---

49. Relis, in interview with LeMenager, quoted in *Living Oil*, 21–22.

CHAPTER 3

"THE HORIZON WAS AN ILLUSION"

Flight, Escape, and Imagining Vertical Space in Leslie Marmon Silko's *Almanac of the Dead*

NATIVE AMERICAN WRITING AND EURO-AMERICAN CULTURE

Two of Leslie Marmon Silko's novels—*Ceremony* (1977) and *Almanac of the Dead* (1991)—inscribe vertical movements and metaphors as an escape from various physical, cultural, and spiritual limits. In *Ceremony*, Tayo is prompted by the sight of the moon to imagine an unassisted vertical journey:

> He had believed that on certain nights, when the moon rose full and wide as a corner of the sky, a person standing on the high sandstone cliff of that mesa could reach the moon. Distances and days existed in themselves then; they all had a story. They were not barriers. If a person wanted to get to the moon, there was a way.[1]

Tayo's imagined flight will erode the "barriers" forged by Euro-American concepts of time ("days") and space ("distance"). The

---

1. Silko, *Ceremony*, 19. All subsequent citations come in the main text in parentheses.

assertion that, under a convergence of the right conditions, a human figure might stand on top of the mesa and "reach the moon" expresses hope in the face of what, in *Ceremony*, is a consistent set of social and legal obstacles left for Native Americans. Refuting the authority of Euro-American modes of knowledge, the passage recalls what Laguna Pueblo writer Paula Gunn Allen describes as the refusal of Native Americans to "fit the categories that have been devised to organize Western intellectual enterprise," just as they do not "fit the cultural descriptions other Americans imposed and impose."[2]

In *Almanac of the Dead* these impositions—and an attempt to move past them—are metaphorically delivered via the physical and cognitive damage left by Root's motorcycle accident. Root knows "how much of his brain came away with the crushed skull." The world has "pulled away and left him lying in white, puffy clouds"; he "could look down and down through the clouds"; he "might have been in a jet airliner except for the silence." As an attempt to escape his body's new limitations, Root mentally detaches himself from his hospital bed, to "look down through layers of clouds and see himself lying in the hospital bed connected to the machines." To imagine he is looking "down and down through the clouds" is a liberation from the constraint of imprisonment in a body that no longer functions as it once did. Imagining flight is also a way for Root to counteract the dreams of "falling forever" that repetitively retrace his memory of the crash during which he fell in a "slow," traumatic descent.[3] In the wider context of the novels from which they come, these two passages mobilize verticality in contrast to the myriad forms of imprisonment enforced by American policies that have sought to dictate the space Native American tribes can occupy, the role they can play in mainstream culture, and the relationship they are permitted with their own cultural history. This chapter proceeds from critical studies in which the specific historical conditions of indigenous people are regarded as essential to an understanding of Native American texts.[4] I read vertical imagery as a reaction to the curtailments of legal and cultural freedom that white America directs against the perceived racial Other.

Few literary works by Native Americans were published before N. Scott Momaday's Pulitzer Prize-winning *House Made of Dawn* (1968).

---

2. Allen, *Off the Reservation*, 6.

3. Silko, *Almanac of the Dead*, 192. All subsequent citations come in the main text in parentheses.

4. I refer here to scholarship by Krupat, Owens, Vizenor, Weaver, Womack, Dennis, and Huhndorf.

In 1983, Kenneth Lincoln proposed that the period after the publication of Momaday's novel marked a "Native American Renaissance" as a flood of literary works emerged by authors such as Leslie Marmon Silko, James Welch, Simon J. Ortiz, and Louise Erdrich.[5] Writing created during this renaissance (and also that following it by authors such as Sherman Alexie and Craig Womack) indicates that the verticality explored by Native American texts is less a fleeting preoccupation than a deeply embedded worldview. In the following, I examine how vertical imagery in *Almanac of the Dead* draws upon themes established not only by Silko's earlier novel, *Ceremony*, but other seminal texts such as D'Arcy McNickle's *The Surrounded* (1936), John Joseph Mathews's *Sundown* (1934) and *Talking to the Moon* (1945), and N. Scott Momaday's *The Way to Rainy Mountain* (1969). The titles of these novels alone reveal coexisting images of imprisonment (Native Americans surrounded on all sides by an often hostile Euro-American presence) and verticality (the sun and moon in the sky and the mountain towering upward).

Almanac is a lengthy, episodic novel that ties together a substantial cast of characters, most of whom are engaged in some form of criminal activity, whether drug-, gun-, or human-trafficking; computer warfare; black market human organ, tissue and fluid donation; animal abuse; "snuff"-movie making; abduction; or murder. Over the course of five parts, the narrative moves from local spaces (Tucson, Arizona), to contiguous nations (Canada and Mexico), to the global (Africa), to "the Americas" (a phrase that does not merely refer to a place on the map but rather an aggregate of indigenous historical, cultural, and spiritual energies), and, finally, a metaphysical and spiritual "fifth world."[6] Studies of the novel predominantly argue that *Almanac* reimagines the historical displacement of native people through the dissolution of borders, borderlands, and Western cartography, and through its narrative treatment of the reclamation of tribal territory.[7] In the model proposed by these studies, *Almanac*'s studied focus on violence echoes the bloodshed of governmental removal policies such as events that occurred on the gold trails

---

5. Lincoln, *Native American Renaissance*.
6. In Hopi Indian and South American Mayan mythology, the "fifth world" refers to the next world. Central to these belief systems is the concept that time is cyclical; the terminus of one world ushers in the beginning of the next. See Allen, *Off the Reservation*, 1; Waters, *Book of the Hopi*.
7. For examinations of *Almanac* and the border paradigm, see Archuleta, "Securing Our Nation's Roads"; Jarman, "Exploring the World"; Romero, "Envisioning"; Tillett, "Price of 'Free' Trade"; Reed, "Toxic Colonialism." See also Beck, *Dirty Wars*, 262ff.

under Cheyenne lands in Colorado in the 1850s and 1860s when Native Americans tried to maintain possession of their land against white gold miners.[8] I read the novel along similar lines while emphasizing an understanding of the "dialectic of simultaneous desire and repulsion" with which the Euro-American greeted the Native American, a phenomenon most commonly referred to as "noble savagery."[9] The double desire to, on the one hand, romanticize Native Americans and, on the other, make them into demonic figures underlies both the specific contradictions of the Indian Removal Bill in 1830 and the more general dilemma of which it was a product: namely, whether to wipe out Native Americans entirely or attempt to assimilate them into Euro-American society.[10] These are the pressures exerted by legal and cultural forces that underlie what Gerald Vizenor calls the "postindian": the Native American as a simultaneous absence and presence in American culture.[11]

In the previous chapter, I argued that Upton Sinclair's *Oil!* reflects how the energies of American expansion, putatively terminated by the closure of the frontier in 1890, were transferred from horizontal movement across the continent into the vertical reach of oil rigs, gold mines, and skyscrapers. This chapter argues that verticality in Native American writing, while also in reaction to the constraints of horizontality, operates with a very different ideology. In the autobiographical *Talking to the Moon,* Mathews watches vultures "hanging against the upper currents, or circling against the blue of the sky"; they are graceful, "with a beauty in flight that is fascinating."[12] The spiritual encounters with the moon and the freedom of natural flight that circulate in Mathews's text are refined by later Native American texts that propose human flight as an escape from the laws of physics as well as from the dictates of Euro-American maps. Moreover, the same tropes are often coupled with allusions to the contested geography of the United States, resulting in a correspondence between representations of the land and an imagined flight that takes place over it. Exemplary is Sherman Alexie's work; as Joshua B. Nelson notes, Alexie's metaphors of "travel through time, space, and all sorts of in-between, ephemeral moments like flight and dancing" work to "reclaim the idea of exploration as resistance against boundaries physical

---

8. See West, *Contested Plains.*
9. Deloria, *Playing Indian,* 3. See also Fiedler, *Return of the Vanishing American;* Keiser, *Indian in American Literature.*
10. See Porter, "Historical and Cultural Contexts."
11. Vizenor, *Fugitive Poses.*
12. Mathews, *Talking to the Moon,* 38.

and imaginative."[13] Where vertical imagery expresses a contract between contested space and flight, it suggests that the land itself holds a latent potency, one which in Mathews's and Alexie's texts can only be released by the recognition that the continent was made up of tribal lands long before European settlement.

In *Almanac,* vertical imagery is counterposed against westward expansion. While representations of the skyscraper and oil rig frequently embody a response to the signaled end of white expansion across the continent, verticality in *Almanac* offers a reaction to the fact that the frontier historically ensured the displacement of an indigenous people who were forced into smaller and smaller territories on the national map. Ultimately, the creation of an American map under white expansion westward not only displaced huge numbers of indigenous people but also renamed the space of the continent through state lines that traversed extant tribal territories. Maps produced in the twentieth century, such as Albert Gallatin's "Carte des Tribes Indiennes de l'Amerique du Nord" (1836), show Native tribal lands before the Euro-American government began to draw state lines over these regions.[14] In contrast to this map, maps from the end of the twentieth century reveal how the tribal estate was literally written over by a new map consisting of state lines. The historical body of these maps reveals a palimpsest in which the subtextual narrative—the ancient tribal lands sensitive to the contours of geographical features such as rivers and mountains—lies beneath the enforced, geometric, and predominantly straight lines of federal states. Coupled with these cartographical documents are myriad legislative ones that participated in the displacement of Native Americans, most famously the 1887 General Allotment Act or Dawes Act, which was named after its progenitor Henry L. Dawes, senator of Massachusetts and chair of the Committee on Indian Affairs.[15] Advertised as a way to assimilate Native Americans by promoting individual land ownership, the Dawes Act abrogated indigenous claims to the land and, by severing the ties of communal possession, foisted Western economic values upon indigenous people.

In his 1899 article, "Have We Failed with the Indian?," Dawes defends the offer of a 160-acre homestead to Native American heads of household, while simultaneously trying to diminish the impact the Dawes Act had

---

13. Nelson, "Fight as Flight," 44.
14. Gallatin's map shows the territories occupied by tribal lands in about 1600 along the Atlantic and about 1800 in the West. The map is held by the David Rumsey Map Collection and is available to view online.
15. For further details, see Trafzer, *As Long as the Grass,* 328–33.

upon existing tribal structures. The "with" in Dawes's title tellingly makes the Native American the object, rather than the subject, of a problem; the question posed by the article is really, "Have we failed to deal adequately with the removal of Native American people?" Claiming that "no English baron has a safer title to his manor than has each Indian to his homestead," Dawes's words emphasize the legal security proffered rather than the questions of what economic solubility (or lack thereof) that same land might afford, and how the "rights, privileges, and immunities of an American citizen" ostensibly accompanying the homestead were outbalanced by the act's detrimental effects upon tribal cohesion and cultural conservation. In fact, the rights and privileges Dawes refers to were only available if the Native American took up residence "separate and apart from his tribe," a caveat revealing that the act's hollow promise of governmental custody was conceived to "fit the Indian for civilization and to absorb him into it." Dawes's essay avoids the underlying question of the legality and morality of a project that rendered the indigenous population "alien" in its own land.[16]

The Dawes Act both rendered the national landscape a disputed space and also secured ongoing dissent from Native American voices. These gush forth via *Almanac*'s depiction of revolutionary cries:

> Only a bastard government
> Occupies stolen land!
> Hey, you barbarian invaders!
> How much longer?
> You think colonialism lasts forever?
> *Res ipsa loquitur!*
> Cloud on title
> Unmerchantable title
> Doubtful title
> Defective title
> Unquiet title
> Unclear title
> Adverse title
> Adverse possession
> Wrongful possession
> Unlawful possession! (714)

---

16. Dawes, "Have We Failed with the Indian?," 280–85; 283. Burt examines how Dawes's essay overtly justifies the Allotment Act, while defending governmental policy as a whole in reaction to criticisms of U.S. foreign policy following the Spanish-American War in 1898, "'Death Beneath,'" 59–62.

These words, called out by the Sioux lawyer and poet Weasel Tail, compound the declarations throughout the novel that Native Americans were "here before maps or quit claims" (216). The Dawes Act is the target of the poem; the "wrongful possession" it refers to is the carving up of reservation land into privately owned portions that forced many Native Americans to sell their lands at a low price to non-Native people. The "barbarian invaders" are those Euro-Americans who had colonized approximately half the original acreage of indigenous tribal land by 1900, constricting tribal territory from nearly 156 million acres in 1881 to around 78 million acres. By 1934 when the Act was finally terminated, the Native American estate had diminished to nearly one-third of its original size.[17]

This emphasis on physical displacement in *Almanac* is accompanied by a violence that is historically evocative of the force with which Native Americans saw their land taken. Not merely have Native American people been subjected to a fraught relationship with the land they once regarded as their own, but within the first century of contact, Native Americans saw the loss of around 70 to 90 percent of their population. As far back as the 1790s, with the Indian Intercourse Acts that established federal control over Native American affairs, the government effectively legitimized crimes against those same people while simultaneously encouraging Native Americans to become small yeomen farmers and thus adopt an Anglo-American way of life.[18] Native American writing is so frequently preoccupied by movement through the American landscape that it is not hard to see the relationship between, on the one hand, a history of forced migration and, on the other, images that express autonomous migration. In spite of efforts to establish a more even standing for indigenous people, the continued displacement of Native Americans across the course of two hundred years has left a legacy difficult, if not impossible, to reverse. The texts I examine in the following chapter provide a response to a history of geographical displacement enforced by cultural and legal forces. Work such as Silko's proposes that by interrogating movement over the American map, and renarrating travel as something that can reach into the sky, it is possible to resist the phantoms haunting Native American tribal history.

The existing, and ostensibly diverse, critical approaches to *Almanac* all focus upon the fact that the novel directs its attention to a historical record of massacre and desecration under the Euro-American colonization of the

---

17. Trafzer, *As Long as the Grass*, 333.
18. See Sheehan, *Seeds of Extinction*; Prucha, *Great Father*.

American continent. In Silko's novel, manifold displacements and forms of deracination, as well as the violence that invariably accompanies them, are the effects of resolutely corrupt Euro-American legal, penal, and governmental procedures. While other studies acknowledge the novel's underlying disillusionment with Euro-American policies, and its concern with the continuing grave situation facing those minorities of the United States caught in cycles of poverty, they fail to address the way in which Silko poses verticality as a response to these problems. While vertical imagery cannot overturn the myriad difficulties facing Native American people as well as impoverished Latino/a, African, Hispanic, Asian, and white Americans, it does condense the desire for the historical roots of these issues to be addressed honestly in the spirit of their future rectification.

"THE HORIZON WAS AN ILLUSION": CLAUSTROPHOBIA AND HORIZONTAL CONSTRAINT

In *Ceremony,* Tayo's uncle purchases Mexican cattle that consistently break through their barbed wire enclosure, prompting Tayo to ask what will happen if they "just keep going, you know, crossing fences all the way back to Mexico." The cattle's insistent migration from their American compound across the national border into Mexico reflects the repetition in Silko's work of indigenous people who face physical, cultural, and spiritual "fences," breaking various borderlines, acting in a way that suggests Native Americans, like the Mexican cattle, have "little regard for fences" (79). In *Almanac,* crossings between the United States and Mexico or Canada are frequent and usually refute legal authority. With a different motivation in each narrative strand—whether tourism, illegal migration, revolution, drug smuggling, or gun running—the border crossings in *Almanac* both rebel against the claims of a Euro-American map and also culminate in events so destructive that they also suggest horizontal motion cannot fully satisfy the underlying urge to roam free.

Almanac takes the spatial circumscription of the national border and that of the Native American reservation and suggests that these both provoke a restless energy that can either be channeled into fruitless—and occasionally lethal—horizontal motion or, alternatively and more productively, verticality. The friction between ideas of horizontality and verticality creates a particularly striking energy in the novel's depiction of David, who escapes the United States to stay with his lover at a ranch in Cartagena, Colombia. At the ranch, dogs and horses are confined in compounds that incite a kind of fatal frenzy when they are then released onto a horizontal expanse. The grooms of the ranch describe the local phenomenon as "rapture of the plain" or "rapture of the wide-open spaces." Local people report "similar strange afflictions" in dogs from the city who are only used to "enclosures"; "unkenneled" on the enormous plains, these dogs "run and run past exhaustion to death" (545). Ultimately both David and the mare he rides succumb to the same fate as they race across the plains to their deaths. The ranch hands have marked David's mare as suffering from "the rapture of the *llano,* the rapture of space and endless horizons." The narrative aligns with David's voice to assert: "David had tried. What more could a man do?," then: "If the horse wanted to run, let it run." David feels enormous "relief and freedom" as he slackens the reins; for a moment the horse hesitates but then she surges forward, "hooves scarcely touching the earth, her sinew and muscle cracking as she raced over the plain toward the horizon's pale blue." The description culminates in the statement: "Truly he had the sensation he was flying" (564). Only when he succumbs to the mare's gallop is David able to feel a release of the tension that has been building in his narrative. Ensnared in a strange love triangle with his manipulative lover Beaufrey and the sociopathic Serlo, David has fled from the complications of his life. Leaving behind his ex-lover Seese, as well as their missing baby, David is distancing himself from lawsuits leveled against him for posthumous photos he has taken of his lover Eric after Eric committed suicide by shooting himself in the head. The horse fulfills David's childhood fantasy of leaving behind "the houses and schools and people" (557). As he realizes this longing, the refrain "if the horse wanted to run, let it run" within the passage replicates the inexhaustible beat of the mare's hooves, creating a rhythm that mirrors the seeming inevitability of her death.

The mare's relentless gallop compounds earlier suggestions that a lack of space results in the release of turbulent energy. Comrade Angelita, one of the leaders of a South American revolutionary army plotting a coup against the U.S. government, reflects upon her army's grievances first privately, and then vocally, and with increasing fervor. Her words announce that when the revolutionaries have taken back the land once owned by indigenous people there will be space for everyone who respects the earth and its inhabitants. She declares that her army is the one to reclaim the land, and it is one army "of many" over the earth. Angelita continues:

> The ancestors' spirits speak in dreams. We wait. We simply wait for the earth's natural forces already set loose, the exploding, fierce energy of all

the dead slaves and dead ancestors haunting the Americas. We prepare, and we wait for the tidal wave of history to sweep us along. People have been asking questions about ideology. Are we *this* or are we *that?* Do we follow Marx? The answer is no! *No* white man politics! *No* white man Marx! No white man religion, no nothing *until we take back this land!* (518, emphasis in original)

Angelita's words imply that the loss of tribal lands has precipitated a kind of human and spiritual pressure cooker that will erupt in an "exploding, fierce energy."[19] This energy is reflected in the phrase, "we wait," which verbally reverberates through Angelita's rhetoric while semantically suggesting an imminent violent fulfillment at the end of this period of waiting. The impending explosion is textually and aurally realized by the climactic string of exclamations prefaced by "no" with a final double negative ("no nothing") that compounds the voiding power of the "European invaders" Angelita is targeting (518). Underlying the speech is the suggestion that the imperative to "take back this land" is both justified and already predetermined by the "tidal wave of history." The same idea is paralleled by the Army of the Homeless, a revolutionary group from the other side of the U.S./Mexico border, led by an African American homeless Vietnam veteran called Clinton. Neither revolution is successful by the novel's close, but to some extent the apocalyptic imagery both groups summon forth is realized, albeit via the bloodshed and disappointment that befalls so many of the novel's characters. Moreover, and as John Beck argues, the "real war" is not "fought by breaking laws that have already been broken," but rather by seeing "the historical connections rendered invisible by the legitimation effects of the nation-state."[20]

Precisely because the violent and implicitly vertical explosion in Angelita's words is precipitated by the loss of tribal land, her speech annexes vertical imagery to a tacit attack on Dawes's 1899 justifications for colonization. Dawes's written defense of his act culminates in the distancing language of the assertion that the seizure of native lands ten years previous was necessary because "there would soon be little unoccupied

---

19. The same sentiment is documented by Frank Waters's interviews with thirty Hopi spokesmen in the 1960s. They offered a prophecy of the imminent coming of the "Fifth World" in which the United States would be overcome by an apocalyptic war, "'a spiritual conflict with material matters [in which] Material matters will be destroyed by spiritual beings who will remain to create one world and one nation under one power, that of the Creator,'" *Book of the Hopi*, 408.

20. Beck, *Dirty Wars*, 265.

room for either race, and it was plain that the two could not live together, and that one must speedily crowd out the other." The passive phrasing of "it very soon became apparent" prefaces the statement that "there would soon be little unoccupied room for either race," making the latter clause seem a fact entirely independent from the actions of the U.S. government. The disassociated syntax of "it very soon became apparent" specifically suggests Dawes's desire to abnegate personal moral responsibility for what was termed the "Indian problem," an idea further supported by his brutal declarations that express a longing for the Native American to simply "fade out of existence in the irresistible march of civilization."[21] In language that makes the subordination of indigenous people appear not merely legitimate but in fact entirely inevitable, Dawes's essay foreshadows the afflictions of a government that could neither justify genocide nor happily accept the jurisdiction of extant Native tribal bodies. By making governmental policies appear beyond the realm of choice—"one *must* speedily crowd out the other" (my emphasis)—Dawes's essay betrays a desire to maintain the appearance of legal validity while forcibly confiscating tribal territory.

The barely veiled dilemma underlying Dawes's treatise is how to exterminate the Indian without recourse to further genocide. The aftermath of the desire to simply erase the Native American, an ambition only further problematized by the Dawes Act's strategies of geographical displacement, was amplified in the following years. Despite the Indian Citizenship Act in 1924, which marked the beginning of a decade of reform, during the 1920s most federal officials were less interested in assimilation than procuring indigenous resources and frequently promoted the continued exclusion of Native Americans from white American society. And, despite Dawes's promise to Native Americans, it was not until 1968 that the American Indian Civil Rights Act allowed Native people the same relationships with local and federal officials enjoyed by other citizens. Mark Rifkin states that the persistence of "official narratives of U.S. jurisdiction," which have "disavowed alternative mappings and sovereignties for too long," "bars" Native Americans from ever "regaining substantive control over their lands."[22] That single word—"bars"—distills the sense that American national policy is best characterized as an incarceration. Feelings of imprisonment that resound in Native texts are also underwritten by the fact that a disproportionately high number of Native

---

21. Dawes, "Have We Failed with the Indian?," 281.
22. Rifkin, *Manifesting America*, 3.

Americans are incarcerated in national prisons, a reality that Luana Ross shows is—like the high numbers of African Americans and Latinos/as serving sentences—the effect of long-term discriminations and ingrained poverty.[23] In *Almanac,* this incarceration is figured as the catalyst for a revenge of the land itself, which is imagined in Angelita's words as an overturning of the map through an apocalyptic scene of destruction.

Dawes's haunting phrase—"one must speedily crowd out the other"— is an image of white imperialism spreading inexorably across the continent like a plague. The same sense of horizontal advance both precipitates Angelita's and Clinton's calls for rectification and is renarrated through the image of David's deathly ride across the Colombian plains. Like Native American people being pushed into smaller and smaller spaces while being robbed of equal rights, the best option for David's previously trapped mare now facing an intoxicatingly open landscape is to express freedom in the only way possible: by running until she can quite literally run no more. Set alongside this is Clinton's apocalyptic imagery, which suggests an imminent explosion from the earth, a time when people are "beginning to sense impending disaster and to see signs all around them— great upheavals of the earth that cracked open mountains and crushed man-made walls" (424). The mare's desperate gallop across the South American landscape is made more futile by these other suggestions in the novel that it will require a transformative geological event to relieve the various constraints placed upon indigenous people.

Almanac challenges the disregard shown to a bloody national past with the urgency that has played out in the critical discourses of the last twenty years calling for a more evenly balanced dialogue between Native and Euro-American histories and discourses.[24] In a manner comparable to its African American counterpart, Native American scholarship often emphasizes the importance of analytical strategies that come from within the cultural parameters of the texts discussed and argues that notions of place, time, and history are differently imagined by indigenous thought in a way that necessarily impacts narrative.[25] As Alfonso Ortiz suggests, spatial ideas such as the border and the frontier, as well as historical precepts such as "assimilation," hold little currency within Native American

---

23. Ross, *Inventing the Savage.*

24. For a genealogy of this critical debate, see Elliott, "Indians, Incorporated." Prominent texts include those by Arnold Krupat, Jace Weaver, Robert Warrior, Craig S. Womack, Gerald Vizenor, Paula Gunn Allen, and Louis Owens.

25. See, for example, Deloria, Jr., *God Is Red,* 62–77. For a reading of *Almanac*'s riposte to Western linearity, see Reed, "Toxic Colonialism," especially 27–28.

communities.²⁶ Nonetheless, and despite calls for a new methodology such as "Native American literary separatism" and then "American Indian literary nationalism," most scholars are less concerned with exclusively Native perspectives and rather continue to insist upon the importance of recognizing the interaction and mutual exchange generated by intersecting Native and European discourses.²⁷ And, although Silko's own affiliation with the Laguna Pueblo filters through the novel's characterization of a Laguna Indian called Sterling, *Almanac* does not lend itself to a tribally specific methodology, rather suggesting that the white drug addict Seese, the exiled Sterling, and the homeless African American Clinton are all equally subject to various forms of social inequality. While Silko's writing often expresses what Joseph Bauerkemper calls "concepts of indigenous nationhood that fundamentally depart from modern state-nationalism and the underpinning ideologies of progressive, linear history," *Almanac* is careful to emphasize the conditions of all who suffer from the lasting effects of prejudice and economic disparity.²⁸

David, too, is neither a Native American nor an African American, but the episode describing his death echoes the desperation in *Almanac*'s depiction of various indigenous and subjugated peoples, and makes his horizontal journey a relentless but unsatisfactory "flight" pitted against the more fruitful images of vertical ascent elsewhere in the novel, which I discuss in detail later. The intoxicating "relief and freedom" David initially feels as he succumbs to the mare's gallop operates in much the same way as the cocaine to which he is addicted. The "rapture" he chases is a mirage; as with the high of a drug, David experiences only a temporary ecstasy before a spectacular crash. The episode is not only a metaphor for drug use, however, but one that culminates in David's death as a kind of sacrifice to the inefficacy of horizontal motion. When the ranch hands find David's body, speculating that the mare "had run until her heart stopped in midstride and she had dropped like a rock," they conclude that he suffered a brutal death: not "thrown free," but caught with the horse as she

---

26. Ortiz, "Indian/White Relations," 1–14.

27. Womack has argued for "the idea of a Native consciousness" while remaining focused on tribal specificity (particularly the unique history and mythology of the Creek people), *Red on Red*, 5. Perhaps unsurprisingly, the apparent extremity of the term "literary separatism" has drawn criticism from those who interpret Womack's as a call for the exclusion of non-Native readers, or assume that his work advocates the maintenance of a Native culture untouched by Euro-American influence. In the more recent *American Indian Literary Nationalism*, edited by Weaver, Womack and Warrior, the editors defend their position in detail.

28. Bauerkemper, "Narrating Nationhood," 28.

fell. Beaufrey and Serlo have to drag the mare off David's body (565). The lure of a horizon that is forever just an unreachable line in the distance proves lethal. The underlying idea chimes with *Ceremony*'s implication that "the horizon was an illusion and the plains extended infinitely" (85). In both novels, the concept of a horizon is tenuous; by extension, movement that directs itself toward a horizon seems doomed from the start.

In the depiction of David's death, *Almanac* transposes the themes of earlier Native American texts that also solicit a sense of claustrophobia in order to amplify an underlying verticality in these narrative ancestors. *Almanac* specifically reframes two preceding texts: N. Scott Momaday's *The Way to Rainy Mountain* (1969) and Zitkala-Sa's "The School Days of an Indian Girl" (1900). Silko seizes upon Momaday's allusions to insanity and transforms them into the emotionally frenzied and physically dangerous horse ride that precipitates David's death. Momaday's text suggests that the Kiowa tribe "reckoned their stature by the distance they could see" and become "bent and blind in the wilderness"; observing a scene in which the earth "unfolds," "the limit of the land recedes," and "the sky is immense beyond all comparison" incites a kind of madness. While comprehending a horizon that is eternally distant makes the spectator "bent and blind," the "immense" sky offers an unlimited "wonder" that is a palliative to the ever-receding horizon.[29] Here the horizontal landscape, even when unpopulated and seemingly endless, feels confining. In *Almanac,* David's fruitless horizontal travel from a difficult situation in North America into an equally dysfunctional world in South America already stimulates his increasing "impatience," and the open landscape he witnesses before his final horse ride incites a "tension" comparable to Momaday's sense of being "bent and blind."

As David looks out at "plains flattened away in every direction until the light blue of the sky folded over them"; the hot air feels like "quicksand" (550–51). The sight of plains "flattened" out, the sky "folded" on top, and breathing in an atmosphere like "sand" combine in the image of David being physically suffocated. In this moment, the text implies that the speed and urgency of David's horse ride is directly caused by a feeling of suffocation, echoing the climactic eruption of horizontal motion in Zitkala-Sa's "The School Days of an Indian." Zitkala-Sa recalls returning to the Dakota plains from one of the many boarding schools that were established under the Dawes Act with the aim of, in Dawes's own terms quoted earlier, making Native Americans "fit" for civilization.

---

29. Momaday, *Way to Rainy Mountain*, 7.

Zitkala-Sa describes the school to which she was removed as a "constant clash of harsh noises, with an undercurrent of many voices murmuring in an unknown tongue"; a "bedlam within which I was securely tied." The imagery of being tied up is exacerbated by the author's suggestion that her "spirit tore itself in struggling for its lost freedom" and culminates in the declaration: "I was again actively testing the chains which tightly bound my individuality like a mummy for burial."[30]

Three years in this boarding school cause Zitkala-Sa to declare, "I seemed to hang in the heart of chaos, beyond the touch or voice of human aid"; physically dislocated, she is also completely cut off from the context of her previous life in Dakota. When she returns to Dakota, Zitkala-Sa describes grabbing the reins of a pony and racing across the landscape, following "the winding road which crawled upward between the bases of little hillocks" where, at the "top of the highest hill," the pony begins "an even race on the level lands." She depicts "nothing moving within that great circular horizon of the Dakota prairies save the tall grasses, over which the wind blew and rolled off in long, shadowy waves." Then Zitkala-Sa asserts, as a paragraph all of its own: "Within this vast wigwam of blue and green I rode reckless and insignificant. It satisfied my small consciousness to see the white foam fly from the pony's mouth."[31] Her imagining of the Dakotan landscape as a "wigwam" restores an indigenous claim upon the land that has been unsettled by the Dawes Act. Meanwhile, the experience of returning home is rendered an explosive confidence in a brief paragraph that is both assured in its declaration, "I rode," and also tempered by the lingering danger of that final qualifier, "I rode *reckless*" (my emphasis).

The school has produced a contained agitation that erupts in a race across and, significantly, up the Dakotan landscape. The hilltop here fulfills the symbolic function of transcending the suffocating "chains" of the federal government. Silko's and Zitkala-Sa's narrations both thematize an agitation set loose on the land, simultaneously evoking the claustrophobic imposition of Euro-American culture upon Native life. However, unlike David's, Zitkala-Sa's horse ride is an expression of freedom that is significant in and of itself: it seems to satisfy the author's desire to once more roam free even if it cannot extinguish the lasting unhappiness caused by her experience of being removed to a federal boarding school. *Almanac*, while also indicating that the theft of tribal lands has produced an

---

30. Zitkala-Sa, "School Days," 186; 190.
31. Ibid., 191.

explosive force that can no longer be contained, poses its horse ride less as a cathartic expression of speed and more as a way to suggest that horizontality cannot free Native Americans precisely because horizontality will always be contained at some point by the limit of the map.

The confinement of the reservation and the expulsion enforced by the confiscation of land under the Dawes Act produce a restless desire for freedom in *Almanac* that is not adequately satisfied by horizontal movement. The closing phrase in the description of David's death ride is "truly he had the sensation he was flying." These words summon a vertical image but do not, at this point in the narrative, fully realize flight: David may have the "sensation" of flying, but he is still moving horizontally. He dies not so much because the flight is only a sensation but because it is directed toward an illusory and unattainable terminus in the distance. While the passing allusion to "flying" is not, here, an adequate release for David, the episode establishes verticality as a contrast to horizontality. It indicates that the claustrophobia of the physically enclosed reservation, the constriction of tribal lands, and the social constraint ensured by government policies have all had a catastrophic effect upon Native people. As much is ratified by recent data collected by Bill Rankin from 2005 to 2010, which reveals suicide rates for Native Americans at their highest level in state reservation areas, where life is segregated from national access to healthcare and levels of poverty, and drug and alcohol addiction are at a national high.[32] The way in which *Almanac* transforms land dispossession into a physically affecting agitation means that the energy of David's deadly ride, and Angelita's and Clinton's violent imagery, cannot quite escape the ongoing painfulness of a geography that makes little sense to indigenous people.

VIOLENCE AT THE BORDER AND THE "IMAGINARY LINES" OF A MAP

Almanac's correlation between constrictions of horizontal space and explosions of vertical motion prefaces the way that the novel yokes the U.S./Mexico border with the idea of attempted escape, and its representation of the border as both a space of multiple crossings and also a place haunted by the same barbarity that accompanied the creation and termination of the frontier. The historical crossings of the U.S./Mexico border indicate the desperation of a people forced to push past the limits of

---

32. See the map of Rankin's findings in Kessler and Jacobs, *Mapping America*, 135.

familiar landscapes in an attempt to regain autonomy. For example, when the government created treaties in an effort to tighten its control over Native American movement in the final years of the frontier, the Apaches in New Mexico and Arizona used the border to try to elude both American and Mexican officials, despite finally being defeated in 1886.[33] The particular brutality directed at Native people in the gold mine regions of Colorado produced by the rush to California is but one example of the correlation between the horizontal conquest of the continent and ensuing violence. Crucially, this violence did not cease with the end of the frontier, and Louis Owens suggests that it is not coincidental that the closure of the frontier coincided with the massacre of nearly three hundred unarmed Native Americans at Wounded Knee by armed troops.[34] The continuing violence perpetrated against Native Americans at the turn of the century, as Euro-Americans traversed so-called "free land," enforced the underlying suggestion in Dawes's essay that describing the Indians as "lawless" justified their displacement.[35]

In *Almanac,* Weasel Tail's zealous cries of indigenous autonomy are mirrored by the assertions of the drug smuggler Calabazas, who claims, "We know where we belong on this earth. We have always moved freely. North-south. East-west. We pay no attention to what isn't real. Imaginary lines. Imaginary minutes and hours" (216). Calabazas's confident assertion that Native Americans pay no heed to mapped lines is belied by the fact that none of the novel's characters are truly able to move freely. Neither the volume of crossings over the U.S./Mexico border by drug runners and illegal migrants, nor the various degrees to which they pose indifference to the law, can undermine the catastrophic effects of these crossings in *Almanac*. In fact, Calabazas's words read less as an assured manifesto and more as the barely veiled longing that guides so much of *Almanac*: the desire to reclaim those lands taken by the Dawes Act and to once more wander free, both of the "imaginary lines" of Euro-American mapped space and the "imaginary minutes and hours" of Western time. A similar sentiment is couched in *Ceremony,* albeit with different emphasis, by Old Betonie the Medicine Man, who states that "Indians wake up every morning of their lives to see the land which was stolen, still there, within reach, its theft being flaunted"; white people "only fool themselves when they think it is theirs"; "the deeds and papers don't mean anything"

---

33. See Thrapp, *The Conquest of Apacheria*; Ball, *Indeh.*
34. Owens, *Mixedblood Messages,* 26.
35. Dawes, "Have We Failed with the Indian?," 281.

(127–28). If the land itself cannot be possessed, the Euro-American map can be made meaningless.

The violence conterminous with the creation of the map, and the entire concept of the frontier as a place for white conquest, continues to undergird the recent history of the U.S./Mexico border. Much has been made of the prescience of *Almanac*'s general rendition of border violence and its specific depiction of a South American uprising by the "Army of Justice and Redistribution" (309), which seemed to predict the actual declaration of war against the Mexican state by the Zapatista Army of National Liberation of Chiapas (South Mexico) in 1994. *Almanac*'s emphasis on the bloodshed resulting from crossings over the border has played out in the last twenty years in the continuing violence of "Amexica," a portmanteau itself indicating the popular conception of the region as a blurry crosshatch of intersecting nationalities. After President Felipe Calderon set the Mexican military against drug producers and dealers in 2006, over 23,000 people were murdered at or near the border before the end of 2010.[36] As well as its dialogue with the contemporary tension generated between American and Mexican agendas, *Almanac* distills the space of the border as an indeterminate and threatening one; in Silko's novel, the border is "the treacherous frontier." Those at its edge find themselves "in the last corner of the United States, the desolate, troubled Southwest territories. There was only direction to go after Tucson; that was *down* to Mexico, and they'd all rather have died first" (644, emphasis in original). Placed as "*down*," Mexico is underneath the Unites States on a horizontal map and represented as a kind of hell even worse than death.

While *Almanac* depicts the border region via transnational crossings that take place aboveground, the many tunnels built in recent years in order to smuggle drugs between California and Mexico suggest that America's international narrative, remaining as urgent as ever, has been forced into subterranean space. On 4 November 2010, Richard Marosi wrote for PalmBeachPost.com that a 1,800-foot passageway built between marijuana warehouses in San Diego and Tijuana had been discovered. A year later on 17 November 2011, Marosi reported for the *Los Angeles Times* that another tunnel had been discovered in the same neighborhood, this time crossing the length of four football fields and coming with

---

36. Vulliamy, *Amexica*, 4. Although "Amexica" is sometimes used pejoratively by conservative politicians, I use it here as Vulliamy does, to indicate how the border has increasingly become defined by the transactions and interchanges that occur over it.

its own lighting and ventilation systems. During its fourth season, Jenji Kohan's cult television series *Weeds* (2005–12) dramatized the creation of a drug and gun smuggling passageway running between a mall shop in South California and a Mexican border town, making the shaft symbolize the hidden presence of drug culture as well as providing the catalyst for its protagonist to increasingly question the broader implications of being involved in that same culture. By contrast, *Almanac* confirms the aerial space over the border as the site of multiple narratives that run counter to U.S. law; the novel reflects that it is "a simple matter" to "avoid the radar along the border" by flying planes "a hundred feet above the mesquite groves"; the drug smuggler Ferro has made it his mission to create new ways of circumventing the border (185), even cruises in hot air balloons that carry tourists along with cocaine and other illegal goods (457). *Almanac*'s representation of the border extends the same underlying correlation between futility and horizontality I discussed earlier, and makes the border so inextricable from violence that even verticality operating over this space is attached to the bloodshed below.

While in *Almanac* vertical imagery is frequently posed in distinction from the desperate violence of the border, it is not immune from the power of the same insidious racism attached to the project of Euro-American colonization in laws such as the Indian Removal Bill. In the most graphic demonstration of this relationship between imperialism and verticality, the novel offers a powerful image of white supremacy through Serlo's exposition of "Alternative Earth module research." The Alternative Earth model echoes the efforts of the Third Reich, is directed at the preservation of aristocratic "*sangre pura*" [blue blood] while exterminating all other races and social classes, and imagines a frightening vertical world. Underground vaults have been constructed to store currency, water, wine, and dehydrated food. An underground chamber called the Alternative Earth unit has been built to contain the "plants, animals, and water necessary to continue independently as long as electricity was generated by the new 'peanut-size' atomic reactors" (542). But these preparations extend beyond disaster scenario preparations; in Serlo's vision of the future, the earth will be left deliberately "uninhabitable" and, at this point,

> the Alternative Earth modules would be loaded with the last of the earth's uncontaminated soil, water, and oxygen and would be launched by immense rockets into high orbits around the earth where sunlight would sustain plants to supply oxygen, as well as food. Alternative Earth

modules would orbit together in colonies, and the select few would continue as they always had, gliding in luxury and ease across polished decks of steel and glass islands where they looked down on earth as they had once gazed down at Rome or Mexico City from luxury penthouses. (542)

The passage continues, asserting that the Alternative Earth modules have been designed "to be self-sufficient, closed systems, capable of remaining cut off from earth for years if necessary while the upheaval and violence threatened those of superior lineage" (542–43). Serlo's description is later affirmed by an eco-warrior, who reveals that the "new enemies" of environmental stability are "the space station and biosphere tycoons" who are "rapidly depleting rare species" so that "the richest people on earth" can "bail out of the pollution and revolutions and retreat to orbiting paradise islands of glass and steel" (728).

Almanac's imagery reiterates the potency of what this book has called the vertical frontier: here, an evocation of imagined racial supremacy made material in actual vertical dominance. Silko's images recall the skyscraper that has been "reared" for the "pleasure" of the ruling classes in Central Park in Upton Sinclair's then-futuristic apocalyptic fiction, *The Millennium: A Comedy of the Year 2000* (1924).[37] The image of hermetically sealed Alternative Earth modules orbiting earth specifically realizes the colonization of an outer space not irrelevantly termed "the final frontier" by popular discourse. The social and physical elevation of a "select few" who look down also condenses the idea of a racial vertical hierarchy, here exponentially multiplied by the startling image of those with "superior lineage," making that distance between themselves and those deemed inferior into a gaping exoatmospheric vacuum.[38] In light of these contexts, *Almanac*'s disturbing image of a wrecked earth left to die as the wealthy orbit on paradise islands suggests that if verticality is an imagining of liberation from oppression, it must also be wrested from the grip of a Euro-American technology determined to conquer any space physically available.

In the images I have discussed, *Almanac* extends Gloria Anzaldua's famous suggestion that the U.S./Mexico border is "*una herida abierta* [an open wound] where the Third World grates against the first and bleeds."[39] By mapping various cultures onto wounded space, the novel

---

37. Sinclair, *Millennium*, 7.
38. See also Adamson for a reading of Serlo and the "biosphere tycoons," "Indigenous Literatures," 151–52.
39. Anzaldua, *Borderlands/La Frontera*, 25.

emphasizes what Rebecca Tillett calls "an increasingly transgressive and transnational perspective" in order to "expose the indeterminacy, fragility, and permeability of borders of all kinds."[40] Five years after the publication of *Almanac,* as the effects of militarization along the U.S./Mexico border were intensifying, Silko described the multiple occasions that she had been detained by border patrol because of her appearance, despite holding a valid Arizona passport. She concluded, "It is no use; borders haven't worked, and they won't work, not now, as the indigenous people of the Americas re-assert their kinship and solidarity with one another."[41] Silko's reflections undermine the U.S./Mexico border as a viable concept and, by extension, challenge the ways in which other spatial models such as the frontier are conceived. The proposal that *Almanac* works to destabilize the entire project of creating borders also finds support in the assertion that the frontier is an enforced Western model incompatible with Native American ways of thinking.[42]

However, the ancillary suggestion to this argument—that the frontier is a fixed idea that will not translate into Native texts—underestimates *Almanac*'s appropriation of the concept in order to reimagine it in vertical terms, in much the same way that Anzaldua's *la frontera* signals the transformation of a Eurocentric model into a theory that recognizes the "contact zone" as a two-way discursive interaction.[43] My line of argument here is illuminated by Louis Owens's argument that the frontier, precisely because it bears the weight of a colonial discourse "grounded in genocide, ethnocide, and half a millennium of determined efforts to erase indigenous peoples from the Americas," is an apt concept for the processes of "appropriation, inversion, and abrogation of authority" with which the Native American trickster tradition is concerned. The frontier, always "a space of extreme contestation," can become the "zone of the trickster," a "shimmering, always changing zone of multifaceted contact with which every utterance is challenged and interrogated." Owens argues that from the very beginning of relations between Native Americans and Europeans, the latter has sought to "inhabit and erase an ever-moving frontier" while "shifting 'Indian' to static and containable 'territory'"—territory here referring to both the static "trope of the noble and vanishing red man" and the "strategy" of elimination that "equated good Indians with dead

---

40. Tillett, "Price of 'Free' Trade," 330.
41. Silko, *Yellow Women,* 108–14; 122.
42. Ortiz, "Indian/White Relations," 2.
43. See Salvidar, *Remapping American Cultural Studies.* The term "contact zone" originates in Pratt's *Imperial Eyes.*

ones." In the face of this historical inequity, Owens argues for a "transvaluation" of "the deadly cliché of colonialism."[44] I extend Owens's main premise to suggest that *Almanac* appropriates the idea of the frontier but, crucially, relocates it to vertical space.

Almanac's insistent portrayal of violence at the border, as well as vertical transgressions of this space, operate as an indictment of the entire project of a map that impairs the freedom of indigenous people. The novel imagines a cartographical taxonomy that has not merely created a violent borderland but has also disassembled the coherence of a Native American "home." Despite the conflicting and complex representations of space in Native American writing, there is one basic tenet concerning the reservation that holds true; namely, that this space, although termed "home" by many, is anything but the original space occupied by Native American people. This is translated into the disillusionment and dissatisfaction associated with the limits of the reservation in novels such as McNickle's *The Surrounded,* Momaday's *House Made of Dawn,* Erdrich's *The Bingo Palace,* and Silko's *Ceremony.* The relationship between confinement and restlessness as a guiding metaphor for Native American experience partly accounts for a bevy of novels that have what William Bevis calls a "homing plot." In these novels, the Native American protagonist travels beyond the reservation only to finally return to its confines. Bevis suggests that this movement forms the "typical Indian plot," which "recoils from a white world in which the mobile Indian individual finds no meaning" and, "as if by instinct, comes home."[45] Where the white American literary tradition has been marked by "lighting out for the territories," in Native American literature "coming home, staying put, contracting, even what we call 'regressing' to a place, a past where one has been before, is not only the primary story, it is a primary mode of knowledge and a primary good."[46] Bevis argues that, through an essential misunderstanding about what this return means in Native American terms, the homing plot may appear futile or negative. Although characters such as *The Surrounded*'s Archilde appear to succumb to "personal doom," their decision to return to the reservation is, in fact, the affirmation of tribal "customs, rituals, and practices

---

44. Owens, *Mixedblood Messages,* 25–27. Owens, like most contemporary Native American theorists including Vizenor, uses "Native American" and "Indian" interchangeably, although makes a distinction between the historical use of the term "Indian" and its reappropriation by Native American people. This idea is extended by Vizenor's suggestion of a difference between "indian" and "Indian," *Fugitive Poses,* 14–15.

45. Bevis, "Native American Novels," 598.

46. Ibid., 581–82.

of law which bind people together into more than a population."[47] Bevis further claims that the contrast between white literary traditions of exploration and conquering ("centrifugal") and indigenous novels of coming home ("centripetal") should not encourage a dichotomy of positive (outward) and negative (inward) movement.[48] In light of these comments, the suffocation that seems to accompany a final return to the reservation in novels such as *The Surrounded* is not the most significant factor at play.

Yet Bevis's model overlooks the claustrophobic themes that accompany movements away from and back to the reservation in Native American texts and severely underplays the emphasis placed upon verticality in contrast to horizontal movement. For example, Bevis's paradigm is strikingly discordant with the visual strategies of Chris Eyre's film *Smoke Signals* (2003), in which the ambivalence associated with traveling beyond the reservation suggests that the "homing plot" acts out its own forms of constraint. In *Smoke Signals,* Thomas's road trip from the impoverished Coeur d'Alene reservation in Idaho and his ultimate return to this space seems a journey that is necessary if the protagonist is to fit into his community, but the ostensible freedom of his travel is undermined by both its slowness and the visual repetition of tropes of confinement throughout the film. As they leave, Thomas and his companion Victor pick up two women in their car, and the ensuing conversation is determined by a feeling of enclosure; the sequence is narrated by flash cuts between the men framed by the car window and the women inside its cramped space. The pervading sense of claustrophobia, even as the two men are escaping the economically stagnant reservation, bleeds out into the world beyond as they travel to Phoenix at a tortuously slow pace. *Smoke Signals* echoes the sense I noted earlier in *Ceremony* that horizontal movement, even when apparently a rebellion against the constraint of Euro-America, is both insufficient and somehow essentially torpid.

Being at home is a tenuous proposition throughout *Almanac*; most characters feel ill at ease in their domestic spaces and the larger communities they inhabit, especially the Laguna Pueblo Indian called Sterling, a "lost and sad" gardener who at the opening of the novel has been banished from his community by the Tribal Council (22). Sterling's exile from, and final return to, the Laguna reservation in *Almanac* challenges Bevis's model by suggesting that the reservation space is haunted by its inception as a prison for the Native American upon the American map. Sterling's banishment,

---

47. Ibid., 583; 586.
48. Ibid., 582. See also Furlan, "Remapping Indian Country," 56.

recounted at the novel's opening, is precipitated by the theft eighty years previous of Laguna "stone idols" that "had accompanied the people on their vast journey from the North" and are old as "the earth herself." The stone figures are eventually rediscovered at a museum in Santa Fe, where a Laguna delegation tells the museum's curator that "these most precious sacred figures had been stolen"; the museum is "in the possession of stolen property" according even to "the white man's own laws." The Laguna delegation meets with a curt response by the curator, who suggests that they "should contact the Indian Bureau or hire a lawyer" (which they cannot afford). This grievance combines with "hundreds of years of blame that needed to be taken by somebody, blame for other similar losses" (31–34).

The theft of the stone idols combines with the opening of a uranium mine on Laguna land—it was 1949, and the United States "needed uranium for the new weaponry"—to produce a metaphor for the various thefts of the U.S. government that have robbed Native Americans of both physical property and a relationship with the earth. The mine destructively channels into the Laguna homestead, "ripping open Mother Earth" near to "the holy place of the emergence," which is a spiritually sacred site (34). Because of the mine, a Hollywood film crew manages to view a sacred object—the giant stone snake—and Sterling, who has been appointed "to keep the Hollywood people under control" (35), is exiled for his mistake (92–98). The triple invasions enacted by Euro-America in Sterling's narrative—the theft of the Laguna people's holy objects, the demolition of their land, and the violation of their spiritual beliefs—form a comprehensive symbol of the myriad infringements imposed on Native Americans. Sterling's home, the reservation, is a stifling space where Euro-America continues to encroach upon Native possession, consistently changing the terms of what the Laguna people can call their own. *Almanac* furthers the recurring theme of suffocation in Native writing by making it intrinsically related to the idea of the U.S. map.

Almanac's suggestion that Sterling's home is merely a prison by another name crystallizes the tension generated by the U.S./Mexico border while also reflecting upon other political acts that make the entire nation into a kind of prison. At its publication, *Almanac* both shadowed the increasingly fraught transactions of the southern border and also anticipated more recent post-9/11 issues surrounding the creation of the United States as a "Fortress Continent." In Naomi Klein's examination of the "growing number of free-market economists, politicians and military strategists" who advocate the creation of "Fortress NAFTA"—a "continental security perimeter" drawn from Mexico's southern to Canada's northern border—the author reveals that the dilemma facing conservative Americans today (who both desire profitable trade terms and also prefer a national policy of racial exclusion and anti-immigration) is essentially the same that faced Dawes a century before. She concludes that the Fortress Continent engineers a "new social hierarchy" in order to "reconcile the seemingly contradictory political priorities" of the post-9/11 era, asking:

> How do you have air-tight borders and still maintain access to cheap labor? How do you expand for trade, and still pander to the anti-immigrant vote? How do you stay open to business, and stay closed to people? Easy: First you expand the perimeter. Then you lock down.[49]

The term *Fortress Continent* connotes a hermetically sealed and impenetrable building, indicating that the United States continues to structure its national policy based on the desire to make the American continent a homogenous economic realm, while also continuing to prevent the Latin American and Native American people who exist in the same space from fully enjoying the profits of its wealth. Klein's use of the phrase "lock down" to describe the closed national border parodies military slang in order to emphasize that America's inhabitants are prisoners by another name.

The Fortress Continent implies that anxieties now circulating in conservative politics are essentially the same as those undergirding Dawes's essay. In the latter half of the nineteenth century, the question Dawes addressed was how the nascent Euro-American population could satisfy its hunger for the continent's potential economic bounty while simultaneously excluding the indigenous population from that same wealth. Now the question is how the United States can maintain its standing as an economic superpower (one resting upon the continued use of illegal migrant workers) while simultaneously closing its borders to the perceived threat of crime (most specifically terrorism) from its neighbors. *Almanac,* which emphasizes the shared mythical and cultural ancestry of indigenous tribes from South America as much as those now living in North America, joins together the late nineteenth-century context with that of the late twentieth and early twenty-first centuries, suggesting the issues at stake now are an extension of those that were first raised under white expansion.

---

49. Klein, "Rise of the Fortress Continent." See also Adams, *Continental Divides;* Huhndorf, *Mapping the Americas;* Tillett, "Price of 'Free' Trade"; and Archuleta, "Securing Our Nation's Roads," especially 114.

The question posed by Dawes and advocates of the Fortress Continent is merely rephrased to accommodate different minorities and essentially wonders how America can continue to redraw its own map in order to subjugate those who would dispute its claim to economic and political sovereignty. *Almanac* reacts to this issue in a way that suggests the very idea of a stable habitat is so completely lost that Native American texts have only one recourse: to imagine a new vertical home above, and therefore independent of, the American map.

THE TRANSLATED FRONTIER: AERIAL HOMES IN NATIVE AMERICAN WRITING

Within the various contexts I have discussed, Sterling's narrative both refutes Bevis's model and suggests that neither horizontally conceived movement nor any circumscribed space on the American map can resist the overwhelming force of Euro-America. In the remaining part of this chapter, I suggest that the dissolution of borders imagined by *Almanac* is not a single subversive act but rather part of the novel's larger project in which verticality is emphasized as an escape from the dictates of horizontally mapped lines. I argue that *Almanac* translates the historical reality of the reservation's and the frontier's horizontal taxonomy into the imagining of vertical space. I extend my reading of Sterling's narrative by placing it in conversation with other Native American writing and propose that *Almanac* images a translated frontier underwritten by the idea that a Native American home has been misrepresented by the historic development of a national map.

In *Almanac*, Sterling's narrative frames the entire novel; the early chapter entitled "Exile" that initiates his story comes full circle at the end of the novel, which closes with Sterling's return to the Laguna people in the chapter "Home" (757–63). In its final moments, the novel asserts that the spirit of the land itself is safe from man's desecrations, which is as close as *Almanac* comes to realizing a reclamation of the land by Native people; Sterling understands that the sacred snake "didn't care" about the destruction wrought by the uranium mine and that "the work of the spirits and prophecies went on regardless"; "humans had desecrated only themselves with the mine, not the earth" (762). But Sterling's return is not entirely happy and, most significantly, even the relief of his realization is tainted by the destruction that accompanies horizontal fatigue elsewhere. Sterling understands that the return of the giant stone snake prophesies "the people" who will come from the south and start an apocalyptic war (763). Sterling has also lost any understanding of what the word "home" signifies: "The taste of the water told him he was home. 'Home.' Even thinking the word made his eyes fill with tears. What was 'home'?" (757). And Sterling returns to the Laguna landscape only to exhibit lost and directionless behavior:

> He had never spent so much time before alone with the earth; he sat below the red sandstone cliffs and watched the high, thin clouds. Far in the distance, he could hear jet airplanes, Interstate 40, and the trains. But Sterling found it was easy to forget that world in the distance; that world was no longer true. (757)

Bevis would perhaps argue that Sterling's return to his reservation is coupled with a recuperative energy through which he is better able to face the demands and pressures of Native life, and can reestablish a relationship with tribal customs. To some extent this is a valid claim, but Sterling is in "shock" (757); he is "haunted" by loss and "blood" and "gunshots" (762). Only the sky above Sterling is a refuge for a "true" world free from all the chaos of the future held by an apocalyptic prophecy of bloodshed; only by viewing the high clouds above can Sterling seem to escape the reality of his spiritual distress.

In *The Surrounded,* Archilde looks "upward to the fleckless sky" and decides that nowhere else "in the world" is there "a sky of such depth and freshness"; he wants "never to forget it, wherever he might be."[50] The possibility held by the sky in *The Surrounded* is extended in Mathews's *Sundown* and *Talking to the Moon,* which make the sky a space onto which desires for escape are projected and betray an emerging preoccupation with flight explored in greater imaginative depth by authors such as Silko. Mathews was born in Pawhuska (now northeast Oklahoma) in 1894, was educated at Oxford University in England, traveled widely, and was a pilot in the First World War as well as a long-serving member of the Osage Tribal Council. *Talking to the Moon,* an autobiographical account of the author's relationship with his tribal lands, was written during the ten-year period after 1932 when Mathews returned to his home of the Oklahoma "blackjacks" of Osage County, an undulating piece of land marked by "three ridges" and a "canyon." In his narrative, Mathews states that his return to Oklahoma is the pursuit of refuge

---

50. McNickle, *Surrounded,* 5.

from the "artificialities and the crowding and elbowing of men in Europe and America," the "clanging steel and the strident sounds of civilization," the "tall buildings and walls," and "neat gardens and geometrical fields." For Mathews, like Sterling in *Almanac,* the vertical reach of the open sky offers an escape from Western civilization and its models of time and space. In the blackjacks, Mathews's experience is shaped not by the dictates of systemized time but by the moon, which governs the seasons and, as the title suggests, is less a distant geological fact than a spiritual presence with which the author can commune.[51]

By affiliating the natural landscape with the Native American protagonist, *Talking to the Moon* suggests verticality as both a trait of Osage County's topography and the means of conversing with the spiritual world held by that same landscape. The Oklahoma sky is an untouched canvas, free of the "strident sounds" and "geometric" shapes of European civilization, just as in *Almanac* the sky above Sterling is an escape from the physical advances and topographical confiscations enacted by Euro-America. In *Ceremony,* Silko writes, "If a person wanted to get to the moon, there was a way," making verticality an alternative to the restraint of a landscape marked by the contestations between Euro-American and Native people. In *Almanac,* the "geometric" lines of European civilization Mathews describes are the same catalyst for images of verticality, only here the primary representation of inflexible Euro-American spatiality is the U.S./Mexico border, rather than "tall buildings" and "geometric fields." Laura Furlan calls *Almanac* the "renarrativizing of the phenomenon of relocation."[52] I argue that Silko's "renarration" works to recuperate Indian tribal coherence not by merely imagining reclaiming horizontal tribal territories but rather by proposing the altogether more surprising, and subversive, relocation of Native American expression into the vertical reach of the sky.

Earlier in this chapter I discussed how, in Native American writing, both the reservation and the chartered space beyond it seem to produce a tiring narrative circularity or a sense of overwhelming fatigue. The claustrophobic tones of many Native texts reflect back upon the historical and cultural conditions indigenous people have endured over the course of white expansion across the North American continent. By setting vertical imagery alongside moments in which this sense of constraint is most acute, *Almanac* proposes flight as an escape, elaborating upon

---

51. Mathews, *Talking to the Moon,* 1–2. All subsequent citations come in the main text in parentheses.

52. Furlan, "Remapping Indian Country," 55.

the suggestion in *Ceremony* that a vertical landscape is potent because it is free from the heady and destructive lure of a phantasmal horizontally mapped line. In *Ceremony,* Tayo thinks that, like the horizon, the sky also looks "far away and uncertain," but "touching the sky" is possible; vertical space is pregnant with the possibility of acquisition whereas the horizon is defined by the fact that it is unobtainable (19). As in *Almanac,* Tayo's reflections on travel to the moon are infinitely preferable to the multiple horizontal journeys made across the course of the novel. *Ceremony* resounds with the idea that distance and temporality are not "barriers" to Native Americans, but rather elements that can be negotiated properly through an understanding of the "story" of "how others before you had gone" (19).

In *Almanac,* a chapter entitled "Flying" compounds the recurring presence of verticality as an expression of escape under two guises: literal flight within airplanes and an imagined ascent that predominantly takes the form of a chemically induced "high." In "Flying," Seese, an impoverished white drug addict who spends the majority of the narrative trying to overcome the clutches of her past while desperately seeking her missing infant son, orders a double shot of Haitian rum in a San Diego airport. As she drinks, Seese experiences "apparitions": a disclosure that immediately precedes a shift to the interior of the airplane Seese has now boarded. Because the apparitions are mentioned immediately before the airplane's takeoff is described, the disembodied high of Seese's hallucinations seems inextricable from her literal bodily elevation. On board the plane, Seese maintains her chemical high by snorting cocaine in the bathroom and ordering "two rum and Cokes." Drugs textually linger in the passage: "Coke," as both the abbreviation of cocaine and of Coca-Cola, recalls that the original recipe for the latter was in fact a cocktail of cocaine and caffeine. Cumulatively, alcohol, caffeinated drinks, and hard drugs become interchangeable through the affiliation of the shared nickname "Coke" and the coexistence of Seese's chemical high with her literal position thousands of feet above the ground (53).

It quickly becomes clear that traveling by airplane fulfills the yearning for escape, whether from drug addiction, poverty, or spiritual disenchantment, which plagues so many of *Almanac*'s characters. Herself in flight, Seese reflects upon her navy pilot father's insatiable taste for aviation:

> Her father used to tease her about going up and never coming down. Just flying and flying forever, so whatever bad weather there was down below, dust storms or even earthquakes, you wouldn't be touched.

Seese's father "described what it felt like flying very high and very fast," where "no earthquakes or dust storms could get him" (54). Then the prose declares that, after her father was shot down, Seese imagined him "flying and flying forever: the aviator's vision of heaven" (54–55). The idea of "heaven" that Seese imagines for her father mirrors her own urge to be in constant motion—notably the same desire that drives David to his death on the plains—and her compulsion to experience life through the transitional space of the airport, a place of "disjointed arrivals and departures" (53).

"Just flying and flying forever," a phrase I take to distill the combined desires of the novel, both syntactically enacts and semantically expresses uninterrupted flight. The languor of the double-syllable "flying" (/flaɪjɪŋ/), with its long middle diphthong (/aɪ/), soars phonetically in a vocal enactment of the word's meaning. The repetition of the word "flying," and the alliterative fricatives in "flying forever," replicate the sense of gliding by making the sentence's structure a closed loop in which flying is both the semantic and lexical constant. At the same time, the phrase adequately summarizes the experience of drug addiction that drives not only Seese but many of *Almanac*'s characters to chase a constant chemical high. Flying—not only physically getting on an airplane but also imagining human flight or experiencing the highs of drug use—is potent here precisely because it appears to promise an escape from feelings of imprisonment or entrapment.

Real aviation and the sensation (or imagination) of flight via drug use become increasingly entwined in the novel via the recurring image of drug-running airplanes that are commissioned to cross the U.S./Mexico border (290). Additionally, the always imminent "crash" experienced as drugs leave the body is mirrored by the novel's repeated mention of plane crashes (157–60; 349; 350–51) and a graphic depiction of a balloon crash in which the basket had caught on fire and "tiny human bodies dangled from ropes as one body fell from high above the Rio Grande" (457). Aviation and altered mental states are made increasingly homogenous, so that verticality can both fulfill the desire for escape and also bleed into the novel's emphasis on drug culture. Seese feels she is "floating free of gravity" (111). A character called Mosca drinks with Root and has "broken free even from the laws of gravity"; he is "flying high with beer foam at both corners of his mouth, nose hairs caked white" (199). At another point, a character called Menardo feels "liquor" as an "invisible, warm wing that lifted him up, out of the reach of the words, where he floated more powerful than any of them" (497). The appeal of verticality at these moments is that they are internally realized: "flying high" on cocaine and beer, or floating out of reach on a "warm wing," represent physical sensations terminated neither by the world outside nor the memories of persecution and injustice that percolate through the novel.

This use of vertical imagery in *Almanac*, which itself refers back to and expands that of earlier texts, has been further complicated by Craig Womack's *Drowning in Fire* (2001) and Sherman Alexie's *Flight* (2007). Both of these novels include Native American protagonists who are capable of either literally flying (as Josh can in Womack's novel) or metaphorically "flying" through time (as both Josh and Alexie's protagonist, Zits, can). Womack describes Josh's ability to travel through time in the same terms as his literal physical flight, suggesting the former to be

> kinda like flying. You didn't want to soar high across oceans and circle the globe, that'd be crazy. You could just get up high enough, brush the treetops, where nobody could hassle you. You might give yourself time to think there, where the sun touches the crowns of trees, the crests of hills, in the evening; that's the place where no one can bother you with everything below you on land.[53]

As if in reaction to the suffocation left by horizontal movement in *Almanac*, *Drowning in Fire* depicts the ability to fly as a successful disassociation from the Euro-American world; Josh experiences journeys to "a place where no one can bother you with everything below." In Alexie's *Flight*, a dissolution of time and space similar to that in *Ceremony* is condensed by the image of a conflated land and sky: "Jimmy flies out over the water, over the great lake, until the blue of the water and the blue of the sky are the same blue. He flies until he cannot see any land."[54] Despite variations in how verticality is expressed, Silko's, Alexie's, and Womack's novels propose flight as unrestrained and liberating: a counterpoint to the limits placed upon a Native American map both by governmentally ordained borders and subsequent physical displacement. Importantly, these texts confirm an emerging intertextual process in Native American narratives; *Almanac* is one point along a continuum of textual voices straining to articulate their aspirations for freedom through verticality.

Specifically, *Almanac* furthers the idea that verticality can secure a precise sense of liberation in its depiction of a triad of characters (Root,

---

53. Womack, *Drowning in Fire*, 67.
54. Alexie, *Flight*, 130.

Trigg, and Max Blue) who imagine flight after suffering accidents that leave them, in varying degrees, entrapped in their injured bodies. Max Blue, a drug kingpin and member of the criminal Blue family, first suffers a military plane crash and then later a shooting that leaves him seriously injured and his Uncle Mike dead. The shooting merely compounds changes in Max that his wife believes to be effects of the earlier plane crash. It is as if he had "died that day in the sand and tumbleweeds next to the runway at Fort Bliss"; he wants to be alone and no longer enjoys sex with his wife. After the shooting, Max moves his family to Arizona, crucially because he is drawn to the "skies" around Tucson. He does not tell his wife, but "after the shooting," the New Jersey skies remind Max "too much of the gray fabric inside a coffin lid." He struggles to explain "the importance of the high, open dome of bright blue sky except to connect the sky with the army plane crash outside El Paso" (350).

Later the narrative repeats the same tropes, stating that "Max feared nothing as long as the sky was open, high overhead, and no low, gray clouds of overcast closed overhead like a coffin lid" (637). Max seeks out the high skies of Arizona not because it reminds him of when he fell to earth in a plane near El Paso, but rather because it reminds him of the realization after the crash that he had survived. He had awakened properly after the crash "into the deep, bright-blue depths of sky all around as if he were flying high above the desert, above the earth." Max cannot forget "the instant he had seen the bright blue of the sky with full El Paso sun." He and the plane navigator survive, but the "old Max" dies in the crash:

> A different Max had somehow pulled himself back into this world, but not completely. He could not rid himself of the sickening fear he felt each time he began to feel drowsy and drop off to sleep. "Dropping off" was so much like dying. (351)

Dropping off to sleep is not merely like dying: it also recalls Max's plane falling out of the sky, an event that is not explicitly described here but hovers at the margins of the narrative and has, in fact, already been represented through the description of several other plane crashes (157–60). The "deep" and "bright-blue" sky reflects back to Max Blue both his own name and the recognition that he is alive, and his survival is characterized as "flying high above the desert." The lingering "sickening fear" attached to the plane crash does not rob flight of its capacity for release, but merely lends even greater power to verticality in the novel; if the sky is threatening, it is precisely because it holds the potential to close off Max's final option for escape from the disenchantment he feels with his life.

Max Blue's relationship with the sky reconfigures the aviation in Mathews's *Sundown,* in which the protagonist Chal, a mixed-blood Osage whose life becomes dominated by alcoholism and disillusionment, finds temporary refuge in flight. Chal warms up the airplane motor, then takes off, and soon the sea is "sparkling" below. Chal thinks it will be fun "to keep flying; keep flying into the distant haze caused by the moonlight." As the plane descends "slowly in a wide spiral," Chal senses "his fingertips tingle" and feels that "the emotion which had dominated him for some time" is "slowly being transformed into action," although the action is "vague." Chal, whose name is short for "Challenge," expresses essentially the same motivations as those who imagine or experience flight in *Almanac*; the assertion that he "thought it might be fun to keep flying; keep flying into the distant haze" even shares rhythmic and semantic qualities with Silko's phrasing, "Just flying and flying forever." Chal flies to feel the "emotion" that has "dominated him for some time" transform into "action"; the "emotion" is the desire to resolve an internal conflict left by both intratribal strife and hostility from white America.[55] Acting out flight is given as the only satisfactory way to transcend these problems. In all the texts I have discussed, the final objective is to maintain flight at all costs as the world below continues to exist: flying until the land cannot be seen anymore in *Flight,* flying endlessly and independent of the sun setting in *Sundown,* flying untouched by the "bother" of other people in *Drowning in Fire,* or flying safe from "earthquakes" and "dust storms" in *Almanac.*

Despite Chal's demise in *Sundown,* at this juncture in the novel flight is a fairly uncomplicated metaphor for autonomous escape from the rules of life below. *Almanac,* however, problematizes the same concept by making death a lingering presence in Max's love for the Tucson sky. Similar complexity attaches itself to Silko's characterization of Root, who is depicted riding his motorcycle when a car turns in front of him and throws him to the concrete, causing "much of his brain" to be lost with his "crushed skull" and leaving him with "a leg that dragged" and "slurred speech" (192; 170). Root is also left with the knowledge of "how heavy the body is as it falls—falls so slowly the mass and weight of it pulls everything down in slow motion." Like Max, Root imagines flight, but for him this

---

55. Mathews, *Sundown,* 220.

is a way to distance himself from the trauma of his accident; imagining he is flying is a way to free himself from the "frustration" of his affected speech and cognitive abilities, as well as the physical limitations of his body (194). Crucially, Root's frustration is condensed in the transformation of his world from upstanding to prostrate. He thinks, "The world was never the same after the accident. Vertical became horizontal" (198). That short final clause—"vertical became horizontal"—betrays the novel's ongoing confidence in verticality as a preferred state in which even the most trapped figures can find a release from imprisonment.

While verticality is not imagined free from the threat of past and future violence in Native American texts, the emphasis on a sacralized or essentially inviolable natural force seems—at least to some extent—to neutralize these threats. Looking up at the sky in *Talking to the Moon* recalls the depiction of height and depth in the natural world as the source of a sublime experience and is met by Mathews's memories from childhood through which he experiences the vertical landscape as something deeply correlated to the autonomy of the natural world:

> I used to watch the distant funnels move wantonly across the prairie from a safe place on the edge of a cave-spotted canyon. One day as a boy, as I sat on my pony at the point of a ridge, I watched a tornado wipe out a little town of the Osage. After the storm, which held me fascinated, I rode down into the valley among the debris. (37)

Mathews observes wind funnels and tornadoes from a "safe distance"; held "fascinated" by extreme weather, he can both watch its path and remain apart from its destruction. This kind of respect for the natural world makes verticality in the texts discussed in this chapter resound with the weight of a long-held relationship between the land and Native American mythology, and confirms Owens's suggestion that the Euro-American has "no past" with the land.[56] The Euro-American oil rigs, gold mines, and the skyscraper of the twentieth century made the American landscape a newly vertical place while provoking the conflicted tones of work such as Sinclair's *Oil!*, which equally extols and laments the technological feats of the fin de siècle. Despite this equivocation the invocation of a sublime experience in the face of the vertical is a keynote of Sinclair's novel.

In the Native American texts examined in this chapter, the sublime is invoked at a moment of verticality to affirm rather than abrogate the

---

56. Owens, *Mixedblood Messages*, 8.

natural world; Silko refigures, in the sublime experience, the entire concept of sublimity as it operates in Western thought. Rather than something held by the land that evokes feelings of fear and awe in the spectator, *Almanac* suggests that the earth's sublime verticality would persist without the presence of the human race. Sterling's closing thoughts assert that the "spirits" carry on "regardless" of human activity; he avers that, "burned and radioactive, with all humans dead, the earth would still be sacred. Man was too insignificant to desecrate her" (762). If the sublimely vertical remains untouched by human intervention, Max reflects that his own "death," which will be "the dark, deep earth that blotted out the light of a vast blue sky Max called life," is not as frightening as it initially appears (353). The loss of his life, in which the earth consumes the sky, parallels the images of a huge "upheaval" of the land elsewhere in the novel and suggests that the destruction wrought by human life is ultimately irrelevant. *Almanac*'s scenes of apocalypse, planes crashing to earth, drug highs, and imagined flights all offer verticality as a transformative force distinct from the influence of topographical confiscation and remappings by Euro-America. The novel represents a traditionally sublime moment but translates it into the outlet for a vertically realized world free from restraint, leaving the sense that vertical imagery is used in full awareness of, but works against the impetus for, the theft of tribal lands and the mistreatment of indigenous Americans.

Native American writing exposes vertical imagery freighted with the wish to intervene in discourses that have historically reduced Native Americans persons to silenced figures, whether through civil and geographical legislation, racist vigilante punishment, or popular culture. A large body of Native American writing responds to various taxonomies of national space and temporality, and specifically alters the historical associations of verticality to the sublime landscape in Euro-American culture. These revisionary acts chime with Gerald Vizenor's description of the appropriation and transformation of epistemic and linguistic structures into resistant signifiers of Native American identity. Throughout, *Almanac* works to recuperate a relationship with an indigenous mythology by entering into dialogue with a broad cultural, historical, and literary frame of reference. Reframing themes and tropes from Native American literature, and mounting a literary response to precise historical conditions in a manner comparable to the work of Ginsberg that is discussed in the following chapter, *Almanac* makes vertical imagery an extension of its announced belief that Native Americans "acknowledge no borders" (a statement written on the map that prefaces the novel).

While images of vertical movement over the U.S./Mexico border often become caught in the violence attending that space, the sky is imagined as a refuge from the Euro-American world that harbors the promise of unpoliced movement, and when flight is characterized as an imagined state, verticality becomes emphatically liberating.

Gloria Anzaldua suggests that "by remaining flexible," the indigenous or mixed-blood person is "able to stretch the psyche horizontally and vertically."[57] The same dilation of an internal world is the desired result for the Native American protagonist who has been subject to multiple contestations and crossings over his or her physical location. In this sense, vertical imagery adheres to Vizenor's definition of survivance as a sometimes "elusive, obscure, and imprecise" current that nonetheless engenders "native presence" over "absence," "nihility," "victimry," "deracination and oblivion."[58] And in *Almanac*, imagined gravity-defying flights and plunges can be conducted without fear of intervention; by proposing both flight and chemical "highs" as impervious to external influence, the novel refracts the hope for a recuperative release from a blood-stained history.

In the previous chapters of this book, "waiting for the sky to fall" articulated a sense of anticipation generated by new epistemologies and technologies, but the Native American texts I have discussed here indicate that verticality can also be temporally bilateral. To some extent, *Little Nemo* and *Oil!* both convey a sense of anxiety generated by the unknown future implications of new science, whereas texts such as *Almanac* articulate a response to a known history. Silko's novel makes vertical imagery projective rather than reflective: rather than exclusively looking back to a long and disturbing history in which a minority figure finds itself the object of oppression, *Almanac* projects itself into an imagined future in which vertical disaster is the effect of a war between Euro-American and indigenous American forces. Ultimately, while Native American texts frequently provide a powerful imagining of the imminent unknown, the kind of verticality emerging from this chapter is one that turns both to the past and the future.

---

57. Anzaldua, *Borderlands/La Frontera*, 101.
58. Vizenor, *Fugitive Poses*, 1–22; idem, "Aesthetics of Survivance," 1.

CHAPTER 4

TERROR/POWER

Allen Ginsberg's Nuclear Poetics
and the Space Race

NUCLEAR TECHNOLOGY AND ALLEN GINSBERG'S "POETICKALL BOMSHELL"

In the early morning of 16 July 1945, scientists and military personnel watched in the high altitude desert of New Mexico as the technology that would detonate the atomic bombs over Nagasaki and Hiroshima a month later was tested. In the days before, scientists had taken wagers about the extent of damage that would be caused by the blast, wondering if it would "ignite the atmosphere," and whether its destruction would span the "entire world" or merely "be confined to New Mexico." Those involved were not quite as cavalier as this anecdote suggests, but the test—codenamed "Trinity" by the scientific director of the Manhattan Project, J. Robert Oppenheimer—was pervaded by uncertainty.[1] As spectators watched from reinforced bunkers 5,000 yards away, protected only by sunscreen and dark glasses, "there was a lighting effect within

---

1. DeGroot, *Bomb*, 56–65.

a radius of 20 miles equal to several suns in midday"; "a huge ball of fire" rose "to a height of over ten thousand feet," and an enormous cloud "surged and billowed upward with tremendous power," reaching "36,000 feet above the ground."[2]

For the military personnel watching and those planners in Washington's War Department, the vertical spectacle signified not global apocalypse but categorical victory with the belief that America, now controlling the most destructive bomb ever conceived, would be equipped to deter any future international threat. This view was initially also shared by the general public after the detonations in Japan; to many the mushroom cloud symbolized "not doom, but progress," and the word "atomic" suggested "modernity, power and technological might." The jingoistic pronouncements of military personnel came along with assurances such as General Groves's that death by radiation was "very pleasant" and the long term risks of exposure were minimal.[3] Another spectator of Trinity, Isidor Rabi, called the blast "a vision which was seen with more than the eye. It was seen to last forever."[4] While the test was a visceral demonstration of destructive technology distinct from visual displays such as the dynamo, both prompt a similar narrative reaction. Rabi's depiction of the detonation specifically recalls Henry Adams's description of the turn of the century's scientific discoveries as a new "supersensual world."[5] Like the x-rays and atoms Adams describes, Rabi names the atomic bomb blast as "seen with more than the eye"; the bomb is a "vision" that, in the act of perception, manufactures a slippage of time and makes the spectator feel as though it might last "forever." In narratives such as Rabi's, the atomic detonations of the mid-century belong in the conceptual realm of the sublime, where the beholder is momentarily taken beyond what would be considered ordinary experience but is not placed in immediate danger of death.[6] While Allen Ginsberg's work similarly annexes the vertical with the sublime, his poetry and prose address the problems of simply describing nuclear technological spectacles as sublime without considering the ramifications of detonating atomic bombs. In chapter 2, I proposed that Sinclair's work attaches verticality to the sublime

---

2. Groves, "Memorandum for the Secretary of War," paragraph 3 of 19.
3. DeGroot, *Bomb*, 272; 281.
4. Rabi, quoted in ibid., 61.
5. Adams, *Education*, 319.
6. See Quinn, "Coteries, Landscape and the Sublime," for an examination of how Ginsberg's writing intersects with the sublime. For the relationship between nuclear detonations and the sublime, see Arensberg, *American Sublime*.

experience in order to call into question the underlying desirability of industrial progress. This chapter extends that discussion, arguing that Ginsberg's work articulates the wonder attending nuclear technology but does not shed an understanding of how that awe might occlude the human suffering caused by the abuse of nuclear science.

Two interrelated frameworks—neither of which is entirely adequate here—have been proposed as a way to consider texts of the atomic age. The first of these is an interdisciplinary model termed *nuclear criticism,* credited as beginning in the early 1980s in response to reemerging Cold War tensions. More than a straightforward analysis of the cultural products of the atomic age, this model uncovers the effect of language and imagery upon the "public experience of nuclear weapons."[7] To some extent, it is a critical paradigm directed at the ostensible absence of nuclear weapons in cultural and intellectual artifacts of the period spanning 1945 to 1985. One significant text, Paul Boyer's study, *By the Bomb's Early Light,* explicitly targets this apparent lacuna, arguing that the reality of the nuclear arms race in fact deeply influenced the consciousness—and thereby the cultural products—of the postwar generation.[8] The second critical framework for atomic culture comprises more recent work that looks back over the last sixty-five years and concludes that we can see distinct periods in which the atom bomb is characterized in particular ways. The authors of *Atomic Culture: How We Learned to Stop Worrying and Love the Bomb* (2004) claim four eras exist under the aegis of "Atomic Culture": Early, High, Late, and Post-Atomic Culture. These periods move from reflecting upon the potential advantages of the fissioned atom as a way of providing new energy, to articulating "concerns and fears" while simultaneously encouraging Americans "to disassociate the devastating potential of nuclear warfare from the realities of everyday life," to vocalizing explicit criticisms prompted by the collapse of support for the government's handling of the Cold War, Vietnam, and the Watergate scandal, and finally, to eliciting a retrospective nostalgia for "our nuclear past."[9] These periods are useful to the extent that they

---

7. Taylor, "Nuclear Pictures and Metapictures," 567. Taylor provides a summary and examination of this methodology. One example of nuclear criticism is Rosenthal's "Nuclear Mushroom Cloud," which provides an intimate account of the ways in which the mushroom cloud has been attached to various meanings, some obvious (phallism and/or the embryo) and some more subtle (the brain), and how these meanings subsequently affect the remediation of what nuclear technology might herald.

8. Boyer, *By the Bomb's Early Light,* xviii.

9. Zeman and Amundson, *Atomic Culture,* 2–5.

delimit the various modes in which nuclear texts operate, but relying on them risks oversimplifying Ginsberg's contribution, which evolves according to the political and social climate of all four of these periods and consistently refutes a one-dimensional reflection on nuclear technology.

Ten years after the Trinity test, Allen Ginsberg read out his poem "Howl" at the Six Gallery in San Francisco. Ginsberg's friend and fellow poet Gary Snyder famously forecast that the reading would be a "Poetickall Bomshell [sic]."[10] Snyder's prophecy would prove correct, and Ginsberg's biographer Barry Miles similarly couches the event and its effects in explosive terms, suggesting that the Six Gallery reading "catalysed the Bay Area literary community" and started a "San Francisco Renaissance."[11] Ginsberg's work is by now staunchly installed in the pantheon of twentieth-century American letters, and the man himself is regarded as the forefather of the Beats, despite the fact that, as Jane Kramer notes, he was more properly a brother than a father.[12] And Ginsberg's role in—as well as his reflections upon—the seismic cultural shifts of 1950s and 1960s America has ensured a wealth of criticism leveled at untangling the particular allure of his poetry. The same enthusiasm for Ginsberg's work has also led to more recent scholarship interested in contesting early conceptions of his poetry—as "confessional," as antinomian—and reframing his work in more precise political and literary terms, with more sensitive examinations of concepts such as the sublime; with a more acute understanding of the influence of William Blake, Walt Whitman, and William Carlos Williams on Ginsberg; and with greater attention to the effects of the poet's 1949 stay in a psychiatric institution.[13] "Howl" continues to exert particular weight; the poem is frequently seen as an adequate representation of what are regarded as the most crucial aspects of Ginsberg's writing and the context in which his work was received.[14] The following chapter challenges this characterization of Ginsberg's writing by focusing on his less-well-known work concerning nuclear technology. I address the poetry by Ginsberg that either overtly or tacitly sets out to analyze the

---

10. Raskin, *American Scream*, 1–3.
11. Miles, *Allen Ginsberg*, 194; 185–217.
12. Kramer, *Allen Ginsberg in America*, x.
13. See Kripal, "Reality against Society"; Hungerford, "Postmodern Supernaturalism"; Svonkin, "Manishevitz and Sake"; Lee, "'Howl' and Other Poems"; Hartman, "Confessional Counterpublics"; Jackson, "Modernist Looking"; Hadda, "Ginsberg in Hospital."
14. For example, Rob Epstein and Jeffrey Freidman's 2010 film *Howl* was originally conceived as a project concerning Ginsberg's entire life and work, but was eventually reduced in scale to focus entirely upon the titular poem itself and the subsequent obscenity trial leveled at Ginsberg's publisher, Lawrence Ferlinghetti.

cultural and social ramifications of nuclear technology on American life from 1945 to the author's death in 1997 after the end of the Cold War.

The prose and poetry in which Ginsberg grapples with atomic culture reveals the extent to which vertical imagery in the latter half of the twentieth century is entwined with, and tries to work through, contradictory ideas about what nuclear technology might herald. The full spectrum of reactions to Trinity, Nagasaki, and Hiroshima included both censure from quarters such as the Beat poets and Native American authors N. Scott Momaday and Leslie Marmon Silko, all of whom were concerned by the uranium mining conducted on Native reservations, and a contemporaneous although incongruent kitsch affection for the atom bomb in popular culture. In the 1950s and 1960s, the mushroom cloud and other symbols of nuclear war emerged not merely in science fiction films and novels where they might be expected, but in a panoply of souvenirs including postcards, songs, drinks, necklaces and earrings, and in the names of companies, racehorses, motels, highways, and bars.[15] The verticality of Ginsberg's writing provides a radical departure from these prevalent images of the atomic bomb by directly responding to the atrocities in Nagasaki and Hiroshima. His work considers the effect of a worldview constructed by the reality of nuclear technology and often either registers unhappiness with, or directly challenges, the government's involvement in both domestic and international affairs from a date much earlier than the start of Late Atomic Culture in 1964.

An obvious example is one of Ginsberg's most frequently cited lines: from "America" (1956), the speaker succinctly commands, "Go fuck yourself with your atom bomb."[16] Less-well-known examples include a 1960 letter to fellow poet Gregory Corso enumerating criticisms of American foreign and domestic policy; the second item on the list denounces Nagasaki by asserting that not only did America detonate the atomic bomb when it was unnecessary, but that this same action was greeted by indifference or applause.[17] A 1960 poem, "Subliminal," parodies nationalist rhetoric that favors acute military defense via a series of exclamatory phrases documenting the ostensible achievements of the postwar period, climaxing in the rhythmic invocation of the least desirable of results, including cancer.[18] In these fragments, Ginsberg not only challenges the

---

15. DeGroot, *Bomb*, 272–74.

16. Ginsberg, *Collected Poems*, 154–56. The poems I refer to in this chapter are from this volume, unless otherwise specified. All subsequent citations come in the main text in parentheses.

17. Ginsberg, *Journals: Early Fifties, Early Sixties*, 137.

18. Ibid., 153–58; 157.

decision to drop the atomic bomb but also demonstrates a willingness to question the government's rhetorical strategies and highlight the ancillary effects of nuclear technology such as disease caused by radiation. While maintaining a distinction between literary and political acts, this chapter extends the critical framework of the previous chapter, which argued that in Leslie Marmon Silko's *Almanac of the Dead,* verticality often expresses a form of protest. Here, I proceed from John Stauffer's proposal that what is at stake here is the way in which "voices of dissent" are "shaped and articulated," and "how forms of protest function as aesthetic, performative, rhetorical, and ideological expressions within culture."[19]

The Beat movement was vocally critical of the nuclear program, but Ginsberg was unusual amongst his peers for his simultaneous critique of the effects of the atomic bomb and his willingness to embrace—even revere—the new technology that enabled it. Repeatedly considering nuclear machinery within the frame of vertical imagery, Ginsberg's writing questions whether the sublime moment that in other narratives is so decisively attached to the various moments coming from the same technology (the launch of rockets as well as atomic detonations) might best be considered in the context of spiritual, and bodily, experience. His work declines to reflect either the nationalism or the touristic affection of popular texts. It also refutes the suggestion by these texts that atomic detonation might purge the past in a manner akin to Richard Slotkin's regeneration through violence, a process that describes myth-making in American culture proceeding from the original colonists' recognition that "an opportunity to regenerate their fortunes, their spirits, and the power of their church and nation" could be achieved via violent means.[20] Ginsberg's work casts the same idea aside, invested instead in defusing imminent violence while retaining the marvel of scientific endeavor.

A "PERSONAL ROCKET": NUCLEAR TECHNOLOGY AND THE SPACE RACE

"POEM Rocket" demonstrates the intersection in Ginsberg's work between the atomic bomb and the technology of which it is a product, imagining a vertical frontier that is both directly linked to new technology

---

19. Stauffer, foreword to *American Protest Literature,* xii. See also the protest literature against the Vietnam War included in the same volume, 481–513.

20. Slotkin, *Regeneration through Violence,* 5.

and also its ancillary implications (171–72). Ostensibly directed at the new space rockets but in fact deeply embedded in a reaction to nuclear technology, "POEM Rocket" both responds to the announcement of the "Space Race" by the U.S. government in a way that was unusual for the period and at the same time converts the vertical language of the Space Race into an expression of private wonder. "POEM Rocket" is dated 4 October 1957, the day that the Soviets launched Sputnik and two years after both the United States and the Soviet Union began building ballistic missiles conceived to launch objects into space. Sputnik effectively began the Space Race—not so much because of the competitiveness of the United States in the face of burgeoning Soviet technology, but because the immediate American response to the launch was to worry about Soviet attack from space. Ultimately it was the great sense of vulnerability in the face of missiles and bombs being launched from the sky that propelled the U.S. government into its own plans to send rockets into space.[21] The synchronicity between the space rocket and the nuclear missile haunts Ginsberg's consideration of the simultaneous wonder and potential destruction afforded by the ability to launch objects into space. The Space Race would soon become intrinsic to a national pride and sense of progress but its origins, and Ginsberg's immediate reaction to those origins, reveal a complex conversation between anxiety and a desire for new technology.

"POEM Rocket" opens with an address to an "Old" moon, then turns to a depiction of the astrological object as "possible," closing the first verse with the suggestion that humankind will make contact with "another life." Combining a disregard for conventional punctuation with a seemingly abstruse succession of images that are characteristic of Ginsberg's work, the opening verse is semantically ambiguous. The first line seems to make the speaker's face a kind of moon in which his eyes are a "new" moon, which could refer either to a new kind of moon in contrast to the opening "Old" moon, or to the lunar phase in which the unilluminated section of the moon faces toward earth and is made barely visible. This latter connotation of the new moon effectively negates the same image the line is involved in describing: if the speaker's eyes are a new moon marked with a "human footprint," and the new moon is invisible, then the subject of the line is also sheltered from view. This lends the poem's opening a sense of impasse that is bolstered by the line that addresses the moon in 1957 as only "possible"; as thus far physically untouched by humankind, it technically occupies a theoretical realm.

---

21. See Nye, *American Technological Sublime*, 225ff.

The desire to put a human on the moon was as yet publicly unannounced, and would not be realized until Neil Armstrong's walk nearly twelve years later. By equating a human face with a lunar landscape marked by footprints, Ginsberg discloses an astute awareness of the significance of the Sputnik launch as the precursor to these later efforts to touch the moon's surface. However, the opening lines are not merely prescient, but accurately distill the underlying urges attached to the national desire to reach the moon. After Sputnik, the moon is no longer a distant astrological object, a sight nebulously reflected back to the drunken watcher upon the surface of a river, but rather seems to harbor the possibility of reaching "another life." Making the moon literally marked by human footsteps seems the underlying allusion in the opening line and would ultimately come to be the aim of the Space Race. But in its next lines, the poem withholds the zeal attached to space travel elsewhere and instead implies that the project to reach the moon is in fact worryingly close to acts of colonization. References to politicians crying and making war, Saturn "slave camps," and "Cuban revolutions" occurring on Mars all make the desired goal of reaching the moon into an act that infects the asserted feeling that "All is possible" with violence.

Like the majority of Ginsberg's free verse poetry, in the absence of metrical regularity "POEM Rocket" establishes a structure based instead on parallelism and a formal accretion in size (here, verse length that moves from four to twelve to eighteen lines) in order to achieve its rhythm. The opening quatrain is met by a twelve-line second verse that both increases the length of the first verse and establishes the repetitive phrase heads of the whole poem. Unlike poems such as "Howl," in which parallelism is more overtly present in the enumerative structure of lines headed with the anaphora "who," "POEM Rocket" uses an isolated unit of repetition via the ploce "O" (occasionally to head a line, occasionally not) in conjunction with the iterative pronoun "I." The effect of these two emphases is to create a dialogue between the addressee—which is variously an astrological object (the moon), the masses (journeyers), or a specific individual (Einstein)—and the "I" that repeatedly heads a verb phrase. By setting up a dialogue between the active speaking voice and its intended audiences (moon, travelers, Einstein), "POEM Rocket" makes the imagined actions of "another life" that close its first verse distinct from the speaker's activity that will come in the next verse.

In the second verse, the overarching image is of a future in which earthly political entities vie with alien life, creating slave camps, enforcing human religions upon various planets, and finally looking to science to create a whole new universe if it were to come to "that" (where "that" euphemistically refers to the exhaustion of all available resources in this universe). The image not only recalls Silko's Alternative Earth modules discussed in the previous chapter, it also accords with the overall tone of *Almanac of the Dead*. In Ginsberg's poetry, the vertical expansion signaled by space travel repeatedly shelters the implication that is attached to it: that the same technology also allows America to drop the most calamitous bombs ever conceived from the sky onto unprepared nations below. "POEM Rocket" creates an analogy between, on the one hand, the technology that facilitates space travel and, on the other, the more aggressive military ventures that are always eventually associated with the human exploration of new territories. As such, the poem suggests that the verticality of both space travel and the bombs enabled by the same technology is a disquieting, not celebratory, symbol of American power.

Other poems from Ginsberg's journals also yoke symbols of imperialism to the theoretical results of moon exploration. In one poem dated 1 November 1960, the opening suggestion lacks an adequate reference point (since America itself is generally termed *the New World*), implying that despite settling on the American continent, the people who now live in that space have somehow overlooked some unnamed but crucial event.[22] By then framing space travel in terms of a perpetual desire to be at the forefront of colonization, Ginsberg makes conquering the moon a continuation of the discovery of the New World. The poem is specifically a critique of the aftermath of political decisions such as that to criminalize the Communist Party, and it annexes decisions of space travel to the military acts of torture and bombing. The poem then reads the Space Race as a reaction to American fears of becoming redundant in the global economy via a seemingly anxious question that asks if America has anything to sell to the universe apart from its history and a collection of scientific formulae.[23] The consternation of this question is more parodic than sincere, suggesting that the ostensible progress denoted by vertical travel is motivated by economic desires similar to those that drove the original pioneers to settle in Jamestown. Overall the poem suggests that the country should be more attuned to its own political guilt than the new wealth that might be located on the moon. In a 1957 collaborative poem written by Allen Ginsberg, Gregory Corso, and Peter Orlovsky titled "Moon Prevention," the poets elaborate upon this suggestion that reaching the

---

22. Ginsberg, *Journals: Early Fifties, Early Sixties*, 158–60.
23. Ibid., 159.

moon is a goal motivated by wealth. The writers claim that the moon in fact belongs to the poet, but it describes how humankind will plunder the moon and other planets for their potential as sources of radioactive material. The desire for nuclear weaponry marries with the desire to reach the moon; this poem implies the unquenchable desire for radium, an element involved in nuclear technology, could result in extensive mining operations in space.[24]

While the second verse of "POEM Rocket" both establishes a contrast between audience and addressor (Einstein/travelers/moon and "I") and also speculates on the aggressions of global governments once the moon has been reached, the last two verses reveal the significance of that contrast by effectively denouncing the actions depicted in the second verse. In the third verse, the voice narrates the speaker's own actions. Gone is the destructive activity of politicians and tycoons, replaced by a direct address that asserts the speaker is engaged in the act of writing a poem for the reader. The present tense of this declaration is complicated by its embellishment into the statement, "I write you a poem long ago," a few lines later. The poem then further locates the act of writing both in the present and the past by returning to the present, disclosing the speaker as denuded and "without identity"; the speaking voice is a voice, or words themselves, and has no more corporeality than lines of "pen mark." While an oscillating temporality is suggested, the determinedly pacific verbs that populate the third verse—such as to *be,* to *write,* and to *dream*—contrast starkly with the shouting and crying referred to in the second verse. As such, the verbs associated with the "I" form defuse those joined with the subjects of the second verse and work to displace the vertical colonization by political and industrial forces imagined in the second verse with the creation of poetry in the third.

The third verse also works as a preface to the transformation of the speaker that is unveiled by the final verse. In lines that connect "I" to the forms made by a pen, the second verse intimates that the speaking voice is the poem itself. The final verse fulfills this oblique suggestion by affirming that poetry should displace the aggressive potential of space travel. Here the poem turns away from the speculative tone of its first two verses and instead makes confident assertions that transform the earlier and more worrying associations of vertical exploration into an affirmation of privately realized wonder, beginning with the assertion of a "personal rocket" being sent up into the "Beyond." The poem's last lines

---

24. Ginsberg, *Journals: Mid-Fifties,* 379–85.

intensify the previous asyndeton by discarding conjunctions as well as any punctuation, while also accelerating the rhythmic pace by increasing the frequency of enjambed lines. The ephemerality of the speaker suggested in the earlier declaration of having no body but "pen mark" is now furthered by the speaker's transformation into "pure thought."

The same bodily evanescence is foreshadowed by the anaphoric "without" that heads three successive lines describing materiality and corporeality being shed. The parallel syntax of these three lines creates a cumulative pattern as the speaker first disburdens himself of materials recognizable as the basic elements of the real space rocket, then the ephemera of human military venture, and lastly selfhood, in the climactic declaration of being "without myself finally." The accelerated rhythm and repetitive pattern of the verse suggest that what is described has a sense of urgency not unlike the increasing pressure of a rocket launch. By mirroring the acceleration of a rocket taking off, the poem more firmly displaces the vertical motion of the spacecraft with the act of reading the poem out loud. Furthermore, by shifting attention to the vocality involved in the act of reading out loud, the poem pointedly emphasizes the body. Rather than signaling a kind of death via the loss of corporeality or consciousness, the last assertion in the list of deficiencies—"without myself"—does not negate the speaker's closing bodily expression, which narrates the act of turning to sleep. The asserted loss of "myself" rather enables the poetic voice's imagination of an accelerating "thought" that flies into space above with instantaneous speed. As such, in its final lines, "POEM Rocket" emphatically replaces the vertical trajectory of the space rocket with human cognizance.

As with Ginsberg's more famous "America," in which the speaker's skepticism in the face of technology is palpable, the overall effect of "POEM Rocket" is to satirize the public's apparently limitless enthusiasm for space travel. The poem's allusion to Saturn "slave camps" evokes the concentration camps of Nazi Germany and the Soviet Gulag, and the reference to "Cuban revolutions" occurring on Mars anticipates Fidel Castro's later revolt (which the United States endorsed) against Fulgencio Batista on 1 January 1959. The poem implies that the most urgent moral and political problems facing mid-century America—how to deal with fascist regimes abroad and communist blocs closer to home—will not be resolved by a successful moon landing. In this regard, the poem rejects the weight invested in new technology and, by contrast, suggests that it is a distraction from the far more pressing global concerns of the postwar period. While still extolling the magnificence of the technology itself, Ginsberg refuses to ignore the fact that this same science is applied to destructive ends by alluding to various forms of human suffering that will not be alleviated by a successful rocket launch.

"POEM Rocket" is concerned with the association of the dynamic sublime that Nye argues is the defining way of experiencing the rocket and the atomic bomb.[25] As I discussed in the introduction to this book, the sublime is a feeling left by the experience of being confronted by a scene of overwhelming impact; the moment is imagined as transformative and, crucially, leaves the viewer at least momentarily unable to comprehend what he or she is viewing. Nye argues for a comparable (although not identical) effect in those viewing the man-made feats of engineering and architecture. It is important to recognize, however, that the essential failure to comprehend what is before one's eyes is to some extent incompatible with the man-made technology Ginsberg depicts, which is inherently explainable by the scientific endeavor that has enabled its existence. That is not to say that the sublime is an irrelevant schema in which to discuss his poetry; rather, the paradox I describe is essential to the way in which Ginsberg's work both describes nuclear technology and space travel as an astonishing feat that pushes against human comprehension and simultaneously acknowledges its scientific explanations. Ginsberg is less concerned with unthinkingly invoking the sublime as he is in disentangling it from its purely technological associations in order to propose science, rather than the destruction that ensues when it is used to detonate nuclear weaponry, as the potential source of rapture.

Rather than channeling the sublime through the spectacle of a rocket launch, Ginsberg locates a transcendent experience in individual consciousness and in an appreciation of the fluids and parts of the human body. Crucially, Ginsberg's emphasis on bodily sensation maintains the sense of the vertical spectacle as a prompt for veneration without losing the connection to reality that is presumably shed during the sublime moment.[26] "POEM Rocket" specifically implies that, more than the general wonder of the human body, thoughts housed by that body should be the subject of the fascination and cultural inquiry, which instead is being directed at the Space Race. The poem translates the vertical motion of the rocket into a "personal rocket" created, as the speaker asserts, to deliver a message "Beyond"; a statement elaborated later in the assertion that the

---

25. Nye, *American Technological Sublime*, 225–56.

26. Justin Quinn discusses this idea further in relation to Ginsberg's poetry, arguing that the poet "realizes that 'everyday reality' is a kind of continuous theatre staged by hegemonic political force" ("Coteries, Landscape and the Sublime," 194).

speaker sends his or her rocket to whichever planet might wait for it. The rocket imagined here contains far more potential than the real rockets that would come to be launched by NASA: it can and will land on "whatever" astrological object welcomes it. The final stanza's declaration that this rocket is made of "amazing chemical," more incredible even than the physical cells of a human body, unites the aggressive space travel documented previously. First, the rocket is a gift sent with neither the expectation of reciprocity nor the implication of future invasion, and, second, the image's entirely new "amazing chemical" made of human thought vies with its unnamed but implied competition—the chemicals involved in launching rockets into space.

The sublime, as Rob Wilson notes, etymologically implies "a movement of the mind not so much *under* as *up to,* or *against,* thresholds of higher power." In this sense, the sublime experience is a "fall upward" that "comprises a threat to self-possession" and "brings about an awesome/terrifying disorientation from the ordinary."[27] Wilson argues that the American sublime turns upon "an *identification* with vast space and power, however imaginary this sense," and forms a literary genre that persists "as New World dialogue between 'Americas' past and future," providing a "purchase" on both diachronic and synchronic "dimensions of collective history."[28] The American sublime in this sense is a preoccupation with space, but space imagined as essentially there, always-already-given. What the dawn of nuclear fission signifies is the loss of this very there-ness. And, paradoxically, nuclear fission conceptually destroys the basis of the American sublime and also consummately initiates the sublime feeling of being beyond the ordinary: nuclear fission potentially heralds the end of all human cognition and physical presence, but precisely because of that it also generates the most extreme version of the sublime as that which is beyond every other experience.

By "devastating the ontological substance of things, rendering objects unstable and resonant with subatomic emptiness," nuclear fission revealed "limits that prove to be liminal." The poetic genre Wilson calls the nuclear sublime emerges as an "admonition/premonition of boundlessness situated in the aura of atomic energy" in reaction to this fracture.[29] In this sense, the discovery of how to split the atom potentially stimulates rapture because what it really means—above and beyond its

---

27. Wilson, *American Sublime,* 170–71 (emphasis in original).
28. Ibid., 20 (emphasis in original); 28.
29. Ibid., 246; 227.

ensuing application to nuclear weaponry—is the entire revision of what was previously scientifically possible. But Ginsberg's writing betrays a deeper concern with the fact that nuclear fission, while generating an unprecedented source of energy so vast that it could only be compared to a god-like power, breaks down atoms into progressively smaller parts and, as such, anatomizes the human body into an even more complex form of ever-diminishing components. Furthermore, Ginsberg's writing does not characterize nuclear fission as a nihilistic manifestation of a yawning epistemic abyss, but rather locates the sublime as an experience relational to the human body. In his poetry, nuclear fission produces a spiritual and sublime moment that in turn facilitates a deeper consideration of the inherent power in that science, without its necessary application to genocidal projects such as the atomic bomb. Unlike other texts concerned with the nuclear sublime, Ginsberg's emphasis on the human body ultimately shapes a response to the fact that splitting the atom effectively invalidates the imagined integrity of our bodies, an idea I discuss in the final part of this chapter as akin to a Buddhist acknowledgement of corporeal impermanence.

BOMBS FALLING AND MUSHROOM CLOUDS RISING

In "POEM Rocket," the movement away from the violence of American military and exploratory ventures indicates a latent affinity for the Buddhist ideology and practice that was increasingly affecting Ginsberg's work by the mid-1960s. At the same time, the American public was becoming more vocally critical of nuclear policy in the wake of the government's handling of the Cold War and Vietnam. Ginsberg and his group of friends comprised a committed force in the antiwar and antinuclear movements that protested against the human and environmental effects of underground mining and nuclear testing. In 1971 Ginsberg would correlate the "end" of history in 1948 with the freedom of expression fundamental to both the concerns of the Beat movement and the legality of political protest. Stating that "everything's gotten so complicated," Ginsberg continues to assert that "history came to an end" in "about 1948," and this moment "opened up into some eternal planet-place where whatever magic takes place is all going to be of our own creation."[30] Ginsberg specifies at another point that 1948 marked not just the end of history,

---

30. Ginsberg, *Allen Verbatim*, 20.

but also the moment when "spontaneous poetries" became the prevailing kind and writers started to favor "open form."[31] Although—or perhaps because—these two comments were spoken by Ginsberg at different locations and at different times, their shared emphasis on 1948 links together a revolution in conceptions of time, which Ginsberg terms "magic," with the lionization of free poetic form. Ginsberg would eventually conclude that "a permanent change in awareness" concerning "the notion of Armageddon apocalypse" was the reigning legacy of the countercultural movement, before which such ideas were "considered eccentric."[32] Ginsberg's emphasis on the convergence between new poetic form and the "end" of "history" is also a response to the paranoia left by the possibility of bombs falling from above in wartime and the reality of the atomic bomb. The way in which poetic form elicits a destabilized sense of time recalls the parity between the formal representation of time and its effects that I outlined in relation to Adams's representation of the dynamo in the introduction to this book. Furthermore, in Ginsberg's work it is as if the enormity of the cultural, social, and technological upheaval caused by the ability to split the atom requires an entirely new mode of poetics, one which reflects back upon the sense of impermanence I have suggested is at the heart of what nuclear fission signifies.

Scott C. Zeman and Michael A. Amundson suggest texts of High Atomic Culture predominantly encouraged a disassociation between the destructiveness of nuclear warfare and everyday life.[33] In the same period, however, Ginsberg cleaves overtly apocalyptic imagery to the reality of the atomic detonations in Japan. In a journal entry from 1954, Ginsberg describes a desolate landscape, reflecting upon the dawn of the atomic bomb and its spectral aftermath.[34] In a short poem from a journal entry in 1959, he suggests that America is a nation determined by its deceit and equates the creation and detonation of the atomic bomb with insanity.[35] A mixed poem and prose piece from 1955 titled "Notes at Work" comprises seven verses that culminate in a description of the hydrogen bomb illuminating a vertical landscape of towers and buildings. Ginsberg invokes an apocalypse that has arrived from Asia, a geographical location named as the place that America first sent its own destructive and explosive force. Ginsberg's depiction of the bomb makes it a god-like entity;

---

31. Ibid., 163.
32. Idem, *Spontaneous Mind*, 463.
33. Zeman and Amundson, *Atomic Culture*, 4.
34. Ginsberg, *Journals: Mid-Fifties*, 18.
35. Idem, *Journals: Early Fifties, Early Sixties*, 112.

the prose refers to a force that exceeds limits of time and space. The declarations crescendo with the invocation of vertical space in the spatial specificity of where the description of a bomb explosion exists: among the height of the skyscrapers.[36] These paragraphs harness a familiar lexicon of the sublime event that exceeds human life—the event described exists beyond human scale and knowledge—and also make the nuclear apocalypse a seeming inevitability secured by the allusion to the Hiroshima and Nagasaki detonations. The narrative voice expresses no surprise that, within the parameters of this history, the very same god-like force appears now as a concussive rip through time itself. In these private reflections, Ginsberg reveals the full extent to which the legacy of the atomic bomb haunts his consciousness in the 1950s and 1960s; in a dream recorded in April 1956, he describes the vertical experience of jets thundering overhead and writes of the anticipative sense that planes carrying atomic bombs will emerge from above.[37] During a trip to Greece in 1961, he writes a poem imagining the destruction of the Acropolis by an atomic bomb.[38] Elsewhere he reflects that he is prepared for any eventuality except the bomb.[39] The influence of the atomic bomb on Ginsberg's creative output is not limited to these briefer allusions and dream fragments, but rather provides a sustained presence in his work and repeatedly interrupts meditations on broader social issues.

One such meditation is a poem entitled "Fighting Society," comprising five verses laid out like a series of staircases. The refrain, framed as a question that asks how it is possible to fight society, runs on from the final line of the first three verses, culminating in the image of detonations, which seems to catalyze the rejoinder provided by the final two stanzas.[40] The poem is typical of Ginsberg not only for its run-on lines and absurdist imagery (most notably the affectionate depiction of Walt Whitman joyously clapping in response to the speaker's expressed intent to fight American society), but for its additive structure, which in the first verse moves from two syllables in the first two lines, to five in the third, then twelve in the fourth, and twenty-two in the fifth. I would argue the poem's form is primarily syllabic rather than typographic—by which I mean that the visual effect of a staircase is the result of increasing line length rather than vice versa; the increasing number of syllables in each subsequent line

---

36. Idem, *Journals: Mid-Fifties*, 124.
37. Ibid., 246–47.
38. Idem, *Journals: Early Fifties, Early Sixties*, 229.
39. Ibid., 184.
40. Idem, *Journals: Mid-Fifties*, 323–24.

is the crucial feature of the poem's form and the typographic arrangement it produces is simply a visual manifestation of the former. Thinking about the concept of fighting society in this way uncovers how the poem's dynamics rest upon an accretion of line length converging with the urgency and force of its titular question. The poem's accumulative form also makes verticality (expressed here in the references to airplanes, bombs, and explosions) its climactic imagery.

The first verse lists four images, which to some extent are all related to the human body. In the opening lines, the invocation of the nose deconstructs the body into an individual component; the reference to gargoyles suggests a grotesque parody or uncanny double of the human form; the image of boxers with broken jaws simultaneously evokes ideas of physical prowess and of physical damage; and the reference to homosexuality suggests bodies at the abstract margins of society. This focus in the first verse segues into the opening image of the second verse, which also alludes to the body but this time a multitude of small ones—infants—and then moves to synecdoches of wider society—cribs, cars, houses, estates, derricks. The third verse combines the human images and industrial metonymies of the first and second verses. The infrastructure and military paraphernalia of the nation preface an unpunctuated litany of hurt body parts, so that the verse combines images of technology and war with single words that succinctly conjure an image of damaged human bodies. The dissection of bodies into parts recalls the idea of nuclear fission as the breaking down of units into ever smaller constituents; the poem makes the general references to military combat into a specific allusion to the way in which nuclear war ensures the most extreme disassembly of the human body ever conceived.

Amid the general images of destruction, the image of explosions precipitates the poem's turn from a weary list of impersonal images to the lighter and more comedic tone of the fourth verse. It is as if the difficulty of answering the poem's questioning refrain of how one might fight society is stalled by the idea of nuclear technology; as if the detonations foreclose the possibility of responding to a fight against American military policy in any meaningful way. I do not mean to suggest that the poem alludes to vertical nuclear warfare only to acknowledge the intractability of trying to resist the force of homogenizing, restrictive, and warmongering society, finally admitting defeat. Although the next lines—that invoke Walt Whitman once more—do not answer the poem's question, they refer to Ginsberg's acknowledged forefather in characteristically whimsical terms, again suggesting a return to poetic narrative as the primary way to address the social problems of the 1950s.

The power attached to poetic form emerges in other writing where Ginsberg implies that poetry can liberate "contemplative individuality" from the oppressive forces of mass media systems. In 1959 Ginsberg suggested that "a crack in the mass consciousness of America" has facilitated a glimpse into an enormous, frightening, and verticalized "national subconscious netherworld" that contains "nerve gases, universal death bombs, malevolent bureaucracies, secret police systems, drugs that open the door to God, ships leaving Earth, unknown chemical terrors, evil dreams at hand." This statement depicts new media as a kind of cultural shell that has homogenized consciousness and hides "a subconscious netherworld" in which all the fears and malevolency of mid-century America percolate. What is most striking—though unsurprising given Ginsberg's belief in the power of poetic form—is that he offers poetry as the primary method of seeing through this "crack." Poetry reveals the individual's private, unique, "secret soul," and San Francisco is the site of a revelatory experience among poets who "glimpse something new through the crack in mass consciousness."[41]

In light of these reflections, the turn to Walt Whitman in "Fighting Society" is more of an answer than it first appears. As in "POEM Rocket," poetry is offered as a counter to the prevailing forces of sinister bureaucracies and a suffocating mass media; "Fighting Society" invokes the approval of Whitman in such an assured way that although the central question recurs in an even more energetic guise in the final stanza's closing line, it now appears to pose less of a dilemma than it did before. The intimacy of the poem's turn to Whitman has a neutralizing effect; by the final verse, what is at stake seems to be less the question of how to successfully fight society and more an affirmation of the very process of fighting itself. Furthermore, the latent sense in "Fighting Society" and "POEM Rocket" that poetry has a unique potency is the entire topic of a poem from 1955, in which Ginsberg makes a conjunction between poetic form and the image of a vertical atomic bomb that might light up space with its flame.[42] Here poetic form is referred to in the opening mention of strophe (which names the first part of the Greek ode) and dithyramb (which is specifically an ancient Greek hymn to Bacchus, god of revelry, and more generally any poem with a similarly impassioned function). The strophe and dithyramb are made analogous to the atomic bomb in a reaffirmation of the suggestion in "POEM Rocket" that the sublime

---

41. Idem, *Deliberate Prose*, 3.
42. Idem, *Journals: Mid-Fifties*, 181.

experience should be rehoused in the private poetic experience as a direct contrast to the external and deeply destructive sublime in other descriptions of the atomic bomb.

The poems examined in the preceding discussion neither confirm Ginsberg's work as an affirmation of the supposed void of atomic themes in American texts, nor provide an example of the discrete categories of Atomic Culture. Moreover, Ginsberg's writing uses verticality to gesture toward the sublime experience while moving away from its original configuration in order to confront the paradoxes at the heart of the sublime: that the scientific advances of the mid-century can both effect a sense of being beyond human comprehension and also be explainable, and that the atomic bomb can be viewed from a distance with comparable safety and yet remain inherently dangerous. His writing also abrades the more visible accounts depicting the atomic bomb in the same period, such as eyewitness accounts of the atomic tests and detonations over Nagasaki and Hiroshima, which reveal Ginsberg's poetry and prose as a complex response to the sublime that is lacking in the documentary and journalistic genre.

Among eyewitness accounts predominantly given by scientists and military personnel, the War Department's spokesperson concerning the atomic bomb, William Laurence, provides the only account from a professional writer. On 9 August 1945, Laurence was on board the plane carrying "Fat Man," the bomb destined for Nagasaki, and watched as it was dropped over the city's industrial valley.[43] Laurence's depiction of Nagasaki retains the attention to color and scale of his depiction of the first bomb but infuses the account with a more tangible urgency, describing being 17,000 feet over the Pacific Ocean, with a view that extends for hundreds of miles and allows the ocean and sky to coalesce into one spherical gesture. Laurence describes being "suspended in infinite space" in the moments before reaching Nagasaki, and an enormous flash of light occurs, followed by a force striking the craft he is in, then an incredible fireball that becomes a tower of fire 10,000 feet high, "shooting skyward with enormous speed." Laurence watches the fire "becoming ever more alive" as it ascends, stating that it is "no longer smoke, or dust, or even a cloud of fire," but "a living thing, a new species of being, born right before our incredulous eyes." Laurence's description of the detonation fulfills the criteria of the sublime, producing a moment in which

---

43. See also Rosenthal, "Nuclear Mushroom Cloud," especially 67–68; DeGroot, *Bomb*, 98–101.

the author is "awe-struck" and, despite the fact that he is expecting the explosion, marking a moment beyond the author's full comprehension. But while Laurence speaks of being awed in the presence of a "giant flash" and "pillar of purple fire," his most emphatic declarations turn attention inward. He reflects, "I am on the inside of that firmament, riding above the giant mountains," "letting myself be suspended in infinite space"; his experience of watching the overwhelming verticality of the bomb's mushroom cloud propels him into solipsistic statements led by "I." His suggestion that "there comes a point where space also swallows time"[44] is superficially comparable with Paul Boyer's suggestion fifty years later that "it is as though the Bomb has become one of those categories of Being, like Space and Time that, according to Kant, are built into the very structure of our minds." But Laurence's comments remain determined by a simile that once more shores up the primacy and specificity of his own experience: the climax is his assertion that it is as though he is the "lone survivor traveling endlessly through interplanetary space."[45]

Despite the very different provenance and genre of Laurence's and Ginsberg's writing, the former's rendition of Nagasaki reveals a central tension in the representation of the atomic bomb that Ginsberg's work is equally intent on addressing. This tension, which I would argue haunts Laurence's account, is generated by the question of victimhood in the face of a sublime moment. Comparably little room is allotted to consider the intended targets of the atomic bomb in Laurence's account, which turns from his prefatory existential musings stimulated by the airplane ride to an abrupt question—"Does one feel pity or compassion for the poor devils about to die?"—followed by a sentence that answers in the negative, via the definitive assertion: "Not when one thinks of Pearl Harbor and of the Death March on Bataan."[46] Providing more consideration for the Japanese victims would have been incompatible with both the intent of a journalistic account and the impetus of a governmental employee such as Laurence. But recalling the sublime moment, while obscuring an opposing perspective from those upon the ground with the intensity of that same experience, brings to the fore the same question Judith Butler posed in the aftermath of 9/11: what, exactly, makes for a "grievable" life? Butler argues for a decentering of first person narrative within a global context, a recognition that loss and vulnerability should be venerated as the proof

---

44. Laurence, *Dawn over Zero*, 231–37.
45. Boyer, *By the Bomb's Early Light*, xx; Lawrence, *Dawn over Zero*, 231–37.
46. Laurence, *Dawn over Zero*, 234.

of our being "socially constituted bodies, attached to others," and ultimately proposes that an equal valuation of life be the defining feature of grief (which, by extension, would secure all life as grievable).[47] The concept of grievable life is precisely what underpins the turn from the atomic bomb's spectacularity in Ginsberg's poetry to a sense of compassion for its victims. This process is, significantly, one that works to accommodate awe without losing sight of the continued reality of human suffering.

Unlike Laurence, Ginsberg witnessed neither the test detonations nor the dropping of bombs over Japan. Yet Ginsberg's imagination of those scenes following detonation, as well as the technology that has brought the bomb about, suggest a great degree of personal investment in, and compassion for, the victims of the bomb. Moreover, the bomb is frequently annexed to more generalized images of apocalypse in poems which further ratify that Ginsberg's work eludes the easy categorization applied to texts of atomic culture. A poem composed in 1960 is pregnant with a meaning made possible by the ambiguity of lines declaring an apocalypse has come and the speaker is alone.[48] These lines can be read in a variety of ways: I am the last remaining citizen who exists across the breadth of the land; I am the only citizen who has been diffused across the land; or a combination of these two formulations. Vacillating between these meanings, the poem leaves the apocalyptic event of its subject undefined. Another poem from 1961 imagines that the skies churn with "cancerous vertigo," distilling Laurence's portrait of a light that penetrates for miles, an earth-shattering blast, an enormous fireball and tower of flames accelerating vertically in just six words. Crucially, the term "cancerous vertigo" condenses two vital components of Ginsberg's discussion of the atomic bomb: the dangerous effects of radiation and the human suffering attending nuclear technology, and a sense of vertical disorder secured by the dropped bomb. The poem also exemplifies the complexity of Ginsberg's terms of description when imagining nuclear technology. The revolving sky bears the ability to possibly remove—or perhaps initiate—death, which is described as being drawn out of the speaker's ear. Joining this ambiguous image is the incongruous and almost oxymoronic final phrase, which grants sentience to an inanimate and intangible gloom.[49]

While the allusions to Nagasaki and Hiroshima are here unspoken, Ginsberg nonetheless invokes the impressive vertical spectacle following

---

47. Butler, *Precarious Life*, 20; 7; 21ff.
48. Ginsberg, *Journals: Early Fifties, Early Sixties*, 171.
49. Ibid., 198.

atomic detonation to correlate this spectacle with the bombings of Japan and what they might signify for the future. In a poem from 1961, the relationship between bomb detonation and its effects upon wider cultural and political structures becomes more explicit. The poem does not merely begin in medias res but opens with a rebuttal to an unheard question; the poem describes a prior attempt to speak in its first lines, before plaintively describing an insane world of those hurtling toward their own annihilation, the loudness of a mass consciousness created by the media, and a clamoring to pull the triggers of bombs, then moving from the abstract reality of the image of being over the Atlantic Ocean directly into the metaphysical suggestion of a Blakean concept of space. The poem opens with the declaration and closes with a series of questions headed by "Who knows . . . ?" Unlike "How can you fight society?," the answer to the closing questions are unheard, existing beyond the limits of the poem's end. As such, the poem's structure is neatly reversed: it opens with the answer to an unheard question and closes with three unanswered questions. This formal loop abets the general sense of fatigue within the poem, which has a repetitive rhythm too, expressed via a fairly regular structure that alternates between slightly longer and shorter syllabic lengths in each line. Crucially, the sense of tiredness prompted by the poem's form joins with the uncertainty of its refrain, "Who knows?," to suggest weariness in the face of a nation seemingly intent to usher in its destruction.[50]

The circularity of this poem meets with the ambiguity of the vertical disorder of the term "cancerous vertigo" and is compounded by Ginsberg's reflection in his journals that the future of the earth is an "upside-down candle"; the wax will melt down and smother the flame that is burning up; it does not matter how the candle is held. These lines of poetry end in a simple declaration that names the atomic bomb.[51] Together, "cancerous vertigo" and the "upside-down candle" join familiar images of vertical apocalypse with the idea that a complete overturning of existing reality will come about if the world's superpowers continue to use nuclear warfare. I have argued that Ginsberg's work does not push away the sense of awe attached to moments of scientific endeavor, but considers them within a frame of the disastrous effect their application can have when they are used as instruments of war. Here the spectacularity of the vertical event is celebrated while a wariness concerning the abuse of technology that can destroy life is retained. As this book

---

50. Ibid., 186–87.
51. Ibid., 170.

has repeatedly argued, where verticality expresses the anxieties of imagined disaster, it also provides the nexus of ideas that propose certain cultural moments as the overturning of a known world—so Art Spiegelman imagines that the falling Twin Towers have ushered in "an upside-down world,"[52] and Upton Sinclair implies that the rapacious oil industry could invert a natural order. Ginsberg's imaginative expression of the atomic bomb as an "upside-down candle" projects the same sense of disorientation through what I argue is an elegant metaphor for mutual assured destruction (MAD) in which nuclear fallout (here the melted wax) will destroy the original act of violence or explosion (the flames) in a never-ending cycle of destruction. The fact that the candle is "upside-down" not only secures the mechanics of the circular destruction depicted, but also alludes to the wider sense in which nuclear technology is popularly conceived as the inversion of an established or natural order.

PLUTONIAN ODE: PERFORMATIVITY AND POETIC NEUTRALIZATION

The remaining part of this chapter discusses Ginsberg's poetry from the late 1970s and its suggestion that the finality of mutual assured destruction can be avoided. In these poems, the author emphasizes nuclear technology as remarkable in and of itself, and celebrates it as just one part of a wider appreciation of human life congruent with his increasing interest in Buddhism. In 1978 Ginsberg and his long-term partner Peter Orlovsky spent the summer in Boulder, Colorado, where Ginsberg had helped his Buddhist teacher Chogyam Trungpa Rinpoche establish the Jack Kerouac School of Disembodied Poetics at Naropa University in 1974. In July of their 1978 visit, Ginsberg and Orlovsky joined friends of the Rocky Flats Truth Force to meditate on the tracks outside the Rockwell Corporation nuclear facility's plutonium bomb trigger factory, succeeding in halting a trainload of waste fissile materials. It was during this summer that Ginsberg produced the collection *Plutonian Ode,* which includes the most overtly critical poems of his work concerning nuclear technology, "Ballade of Poisons," "Nagasaki Days," and "Plutonian Ode."

In "Nagasaki Days," the author draws explicit attention to the poem's chronologically additive form across the summer of 1978 by including within the poem notations that the first part was written on 18 June, the second and third on 9 August, the fourth on 10 August, the fifth on

---

52. Spiegelman, *In the Shadow,* 7.

17 August, and the sixth simply over the summer (707–9). The poem's parts narrate the protest at the Rockwell bomb trigger factory; it moves from the first three parts, in which images of a tranquil summer day harbor threatening allusions to nuclear detonation, to a fourth part that steps away from the present and speculates more generally on the act of bombing. Part five of the poem returns to the protest, and the climactic part simply lists figures and in doing so annexes the millions killed in Vietnam with the millions of refugees in Indochina, making the casualties of American military intervention congruent with the nuclear technology the entire poem is protesting against. The fourth section of "Nagasaki Days" imagines a bombing in the United States, possibly in the aftermath of Nagasaki, in which bombed victims retaliate by dropping radioactive material over New York. The poem extends the same speculative narration of nuclear apocalypse that is an iterative theme in Ginsberg's journals, but the context of this poem and its function as a narration of Ginsberg's physical protest against nuclear technology in 1978 makes the apocalypse described congruent with the real scenes left by the bombing in Japan. With the overt signal of its title securing the following lines' conjunction with Nagasaki, the speaker of the poem describes "the bomb'd" dropping "lots of plutonium" over New York's Lower East Side. The poem continues to describe a scene in which no buildings exist, only "iron skeletons," and there are people "starving and crawling" (708).

Ginsberg bravely addresses the reality of a fear that is excluded by Laurence's description of Nagasaki and the initial reports of the Japanese atomic bombs in the American press. In Laurence's description the movement toward considering whether the author (and, by extension, the American public) should feel compassion for the Japanese victims engenders a prompt foreclosure of the question that invokes Pearl Harbor as the justification for Hiroshima and Nagasaki. The swiftness with which this dilemma is expelled from Laurence's prose absents an understanding of what it leads to when considered more carefully—namely, the following question: if the United States drops an atomic bomb in wartime, surely other nations may follow suit in the future? While later texts would also consider this notion, Ginsberg's explicit reframing of the atomic bomb in New York is a direct response to Nagasaki that shifts attention to the vulnerability of the human body. Moreover, "Nagasaki Days" reins in the relationship of space travel to the atomic bomb I discussed earlier, as well as closing with a symbol of more general human pollution and war, "Charred Amazon palmtrees": an image alluding all at once to nuclear apocalypse, the destruction of the rainforest, and more obliquely, the napalm bombing of similar tropical trees in Vietnam. Here the dropping of bombs is made congruent with other acts of imperialist venture in such a way that a specific question—why does the American government fail to acknowledge the intersection of these atrocities?—is brought to the fore.

In "Ballade of Poisons," the link between nuclear technology and its pollutive effects is the primary determiner of both the poem's form and its subject matter (700). This poem adopts the ballade supreme form (three ten-line stanzas, each with the rhyming pattern ABABBCCDCD) with a five-line postscript, offering in pentameter a structured contemplation of radioactivity and the effluvia left by nuclear technology. The ballad form, historically affiliated with the oral tradition, conventionally has a simple narrative theme that is joined by a fairly uncomplicated and accessible structure. These associations are subverted by Ginsberg's choice of subject matter, the bleakness and apparent uncontrollability of which abut the predictability of the poem's form. Not only adhering to the metrical and rhyming scheme of the ballade supreme, "Ballade of Poisons" also constructs another level of regularity in the anaphoric head ("with") of each line. Crucially, each verse is constructed around an anastrophe (the figure of speech in which the conventional word order of English is inverted) as the refrain that closes each verse would ordinarily precede the prepositional phrases of which the first nine lines are composed. By delaying the complementary clause to the first nine descriptor phrases headed by the adpositional "with" preceding the object (including "oil," "soot," "new plutoniums," and "pesticides"), the first verse's climactic lines are only made into a meaningful sentence once the refrain is revealed. As such, the poem generates a particular tension through its structural delay.

The tension is relieved by the last line, which, as the only line in trochaic pentameter, provides a syllabic beat lacking from the irregular pentameter of all the poem's other lines. This rhythmic beat makes the final line phonically emphatic as well as underscored within the poem's overarching structure as the rejoinder to all the preceding lines. Furthermore, because the poem is so pointedly structured, when it deviates from its established form in the eighth line of its first verse, the effect is a clear emphasis on what that line contains. The seventh and eighth lines are not only emphasized by the turn from the anaphoric "with" to "their" but also because their enjambment is unique in the verse; the "new plutoniums" that form the topic of the seventh line is the only subject afforded a line that breaks across the self-contained lines evident elsewhere. What the line describes—a poison so overwhelming that it takes 250,000 years to become harmless—equally exceeds the parameters of its comparably pedestrian counterparts of "soot" and "oil," so that the poem achieves a neat parity between tenor and vehicle.

The overall effect of "Ballade of Poisons" is to marry the less extreme pollutions of man's industrial and technological ventures with the overwhelming detriment caused by abusing nuclear materials. The underlying relationship between poetry and living—here, the unity between poetic form and what it describes—condenses the suggestion that poetry itself is a kind of antidote to the destructive forces of the inhabited world in "POEM Rocket" and the fragments I discussed earlier. The emphasized final refrain of "Ballade of Poisons" has a sufficiently ambiguous syntax as to make its declarations signal either alarm or optimism; its intent can be paraphrased as either, "May you be able to accommodate all these poisons and still be spiritually at ease" or, and more likely given the allusion to tears, "May you be forced to live with these poisons until you fully understand their implications." In either case, the poem extends the immediate effects of nuclear technology (namely the atomic bomb) into its equally catastrophic long-term effects, a process that reminds its audience of the fact that the victims are not merely those abroad during war, but the general American public.

Like "POEM Rocket," which I have suggested is the first of Ginsberg's published poems to directly correlate the Space Race with the atomic bomb, "Plutonian Ode" (710–13) contests the idea that vertical technology should either be greeted solely with rapture or opprobrium. Instead, this poem interrogates the relationship between scientific venture and spiritual transcendence, returning once more to central concerns of Ginsberg's poetry: the human body, the human breath, and the primacy of the poet. With a similar structure to "Howl," "Plutonian Ode" is structured as a series of long lines of free verse set into three parts. Despite the uniformity of its line length throughout, the rhythms, imagery, emphases, and address of each part are very different. The first verse addresses a nonspecified listener, documenting what it calls a "new element" in a circumlocutory manner that never in fact directly invokes the element's name. Instead, we have the poem's title, which simply suggests that the poem will be an ode to matters Plutonian, while recalling the shared etymology of the chemical element plutonium and the planet Pluto, both of which were named after the Roman god of the underworld. It quickly becomes apparent that plutonium is the poem's real topic: that "new element" unnamed but alluded to in the mention of Doctor Seaborg (the scientist who is credited with discovering plutonium) and the suggestion that its name comes from the planet Pluto.

Plutonium is the heaviest primordial element (that is, the heaviest of the elements that have existed in their current form since before the earth's creation) and the byproduct of fission in nuclear reactors. Plutonium-239 is the particular isotope used in nuclear weapons; it is fissile, which means its atoms can split to create a nuclear chain reaction. The references that populate the first verse of "Plutonian Ode" annex the classical mythology of Pluto to the chemical element that takes its name; the poem declares that plutonium is named after Hades, who kept Demeter's daughter Persephone underground for half the year. The closing lines of the first verse provide the image that Ginsberg notes was the inspiration for the entire poem, with their reference to the Great Year, the astronomical term for the period of around 25,800 years required for a complete cycle of equinoxes around the ecliptic. Ginsberg remarks how he was struck by the concurrence between this astronomical phenomenon and the half-life (the time span it takes for a decaying substance to halve in radioactivity) of plutonium, which equals roughly the same amount of time as the Great Year.[53] In the first verse, both plutonium itself (the "new element" that is in fact only new in terms of its human application as nuclear weaponry) and the mythology from which it takes its name are depicted reaching back to before the 167,000 cycles (4 billion years) of the earth's existence. By rooting plutonium's name in classical mythology and paralleling one of its chemical properties with the course of the earth's axial turns, the poem's first verse undermines the line with which it began, a line that describes a new element "unborn." The implication is that plutonium, the primordial element that existed before civilization, has an inherent and age-old potency. While plutonium is not truly a new thing, the first verse argues that it has been reborn—or, rather, "unborn"—as the material that was discovered by Seaborg and subsequently used to construct nuclear weapons.

While this preface distinguishes between plutonium as an element that in itself has existed for millennia and as an element that, in human hands, holds the most abhorrent power, the poem's second verse turns its address to plutonium itself, opening with a question that asks whether this "Radioactive Nemesis" was present at the beginning of the universe. At this point, the first person enters into the poetic voice, a moment

---

53. Ginsberg, "Notes to *Plutonian Ode*," 803–5. In 1999, the world's "first permanent storage facility for nuclear waste" was constructed in Carlsbad, New Mexico. Around the year 2035, the hole will be filled and then sealed, and must be kept shut for "a legislated period of 10,000 years"—until the year 12035 (Van Wyck, "American Monument," 149–50). This period, though itself almost unimaginable, is only 4.16% of the full period required for plutonium to decay to safe levels (240,000 years).

emphasized by the parallelism of present tense noun phrases headed by the first person, including "I begin" and "I yell." The repetition of this structure couples with the immediacy of the present tense in such a way that the verse is a forceful rendition of the speaker's active response to the "Radioactive Nemesis" it addresses. Importantly, the verbs appended to the "I" repeatedly correspond with speech—chanting, yelling, vocalizing, roaring, speaking—in order to foreshadow the poem's later emphasis on Buddhist chanting and the invocation of poetry as strategies of protest against nuclear technology.

Like "Fighting Society," "Plutonian Ode" invokes Walt Whitman, and even intratextually reconstructs "Song of Myself" by rearranging its crucial assertion, "I celebrate myself," with reference to a "matter" that effects "Self oblivion." The "celebration" of plutonium documented here, and throughout, is crucially both full of awe and cynical. The poem is emphatically concerned with the contrast between the power immanent in radioactive material and the desolation effected by humanity, which harnesses plutonium in nuclear reactors, producing "death stuff" and the decay of nuclear materials. Yet the poem's early allusions to classical mythology, which are joined by mention of various other religions and bolstered by the pervasive references to Buddhism, correlate the poem's subject—a matter that "annihilates" the material objects of writing and prayer—with the omnipotence of various gods. The increasing urgency of these declarations reaches a crescendo in part I with several allusions to Buddhist practice. The "six worlds" that the speaker describes as vocalizing plutonium's "consciousness" denote the Buddhist concept of the six worlds of "Gods, Warrior Demons, Humans, Hungry Ghosts, Animals, and Hell Beings" who are, in Ginsberg's own words, "held together in the delusion of time by pride, anger and ignorance."[54] The assertion, "I chant your absolute Vanity," alludes to Buddhist chanting and foreshadows the final lines of the poem that, as I discuss below, paraphrase the Sanskrit Prajnaparamita mantra. These lines deploy the language of Buddhist thought as a way to articulate the extent of the power held by plutonium that has been roused by the nuclear testing occurring in plants across the country. In repeated exclamations, the poem suggests that this same process has transformed an inherently "Ignorant" material into an "unnatural" force of destruction.

The apparent incongruity between the poem's awe in the face of what plutonium signifies and its simultaneous abhorrence of the annihilation

---

54. Ginsberg, "Notes to *Plutonian Ode*," 805.

it might produce is less an unresolved conflict and more part of an expressed attempt to defuse the "detonative" possibility of plutonium. Crucially, it is not plutonium itself but rather the process that renders it an "Incinerator" of human life that is the target of the poem's hostility. Furthermore, by its close the poem has imagined the swallowing up of the volatile nuclear material it describes. Having reached a climax in the repetitive exclamatory assertions stated previously, the last section of part I slows down once more, occasionally making shorter exclamations but now largely concerned with narrating a series of actions in which the speaker enters the underground realm of the plutonium store. In this section, the self-reflexivity touched upon previously with the phrase, "I begin your chant," comes full circle, as the speaker asserts an act of inscribing "this verse" as a prophecy with the intent of sealing the "you" up, "Eternally" and with "Diamond Truth." In the earlier line, the poem describes both its initiation and its vocal enactment as the speaker narrates that he is beginning the poem's "chant" while the poem is in fact already in progress. The later lines go further, suggesting that as the poem is being created and read out, it is also simultaneously being written down in physical form: sketched out upon the funereal walls of the nuclear plant.

Once more, Ginsberg makes Buddhist practice essential to the poem's performative gestures of speech and inscription, as well as its enactment of nuclear pacification. The Diamond Truth, part of the Sunyata doctrine that depicts existence as "simultaneously void and solid, empty and real, all-penetrating egoless (empty void) nature symbolized by adamantine Vajra or Diamond Sceptre," is invoked as the way to "seal" plutonium up forever.[55] The implication is that the act of reading out the poem combines with the writing of it on the walls of the reactor's core to imprison plutonium; this prison is a vacuum in which radioactivity is eternally suspended between the being and unbeing manifest in the Diamond Truth doctrine. The desire is that plutonium retain the wonder of its inherent power without the risk of it being used to effect mass destruction. Rather than negate the context of plutonium as a material used in warfare, what Ginsberg seems to propose is akin to the process enacted by the Buddhist Heart Sutra, in which the bodhisattva embraces what would otherwise be considered a paradox of oppositions and instead recognizes that both elements involved are equally "empty."

The final words of part I of "Plutonian Ode," "O doomed Plutonium," fulfill the missing address of the poem's beginning, calling the

---

55. Ibid.

material by its name for the first time after forty-four lines of various allusions. After the self-contained (since it ends by answering the unnamed subject of its opening) and first-person address of the first part, part II moves into the third person in order to objectively narrate the scene in which the first part has been spoken. As I have noted, Ginsberg was in Boulder, Colorado, when the poem was composed; the first lines of part II draw attention to the specific time and place of poetic creation in the description of "The Bard" surveying plutonium's history and contemplating "Satanic industries" the world over while physically in Boulder, Colorado. The repeated ellipses in these lines, most usually of the definite article, combine with their paratactic syntax to produce a fast, breathless rhythm and also evoke a sense of detachment. Combined with the third-person point of view, in which the speaker is both "The Bard" himself and is seen to observe that same person, the second part of "Plutonian Ode" has a far more dispassionate feel to its predecessor. As such, it moves beyond the enactment of the poetic line that in part I is expressed via those previously noted moments of self-reflexivity in order to locate the specific time and place of the poem's conception.

If part II is a reflection back on the poet, part III expands the metatextuality of part I in order to reflect further on the poem itself, addressing future poets and orators with lines that command the addressers to "take" the poem in various ways. After these imperatives, the poem climaxes with a final line that Ginsberg notes is an "Americanese approximation and paraphrase of Sanskrit Prajnaparamita (Highest Perfect Wisdom) Mantra: Gate Gate Paragate Parasamgate Bodhi Svaha."[56] This mantra, also known as the Heart Sutra, is a central tenet in the Buddhist religion directed at an understanding of the human mind as a kind of chamber for the infinitude of the universe. Its central teaching comes from the idea that what we conceive of as the mind is made up of five skandhas—form, feelings, perceptions, impulses, and consciousness—that are constantly mutating energies, and that our attachment to these skandhas is the source of our suffering. Only by properly attaining freedom from the skandhas—via the recognition that they are "empty"—can one be free from suffering. This concept of "emptiness" informs the process of reaching nirvana and is echoed in Ginsberg's suggestion of a vacuum in which plutonium is suspended in a permanently powerful state and cannot actually be detonated.

The Buddhist concept of nirvana accommodates neither contradictions nor pairs of opposites but rather a sense of stillness, while the Heart

---

56. Ibid.

Sutra teaches that using the experience of emptiness to find truth enables an enlightenment beyond even nirvana. The last lines of the Sutra Ginsberg paraphrases in the closing lines to "Plutonian Ode" speak of the function of enlightenment in the world; it is a call for the application of truth through helping others. Where the Heart Sutra operates a mode of describing via negation—shedding the restraint of the skandhas is articulated as follows: "All things are empty appearances. They are not born, not destroyed, not stained, not pure"—to suggest that the attainment of wisdom abrogates the need for speech, the climax of the process enacted in the Sutra (truth) is about acting out enlightenment through compassion for others.[57] Additionally, the Sutra itself functions as a way of acting out—of performing—words with the intention of focusing the mind toward a particular goal.

The same can be said for Ginsberg's poetry, which appropriates the closing line of the Heart Sutra as the culmination of its driving force, locating the spiritual experience that can be garnered from the wonders of scientific progress within the prosaic, but no less celebrated, expulsion of human breath. In "Plutonian Ode," early moments of metatextuality—when the poem refers to the vocal "chant" and then the inscription of its words on the walls of the nuclear bunker—meet with the final expression of Buddhist chant to make the poem into a balm that tries to assuage the violent explosions of plutonium with the soft ones of human breath. Finally, the poem bids, "Enrich this Plutonian Ode to explode," both referring to its own title and also self-reflexively creating a phonetic explosion through the only internal rhyme of the entire poem in "ode/explode." The Buddhist Diamond Truth alluded to earlier becomes clearer as the poetic voice expresses a desire for the poem itself to erupt in "empty thunder" (a neat image of both presence and absence) and eradicate plutonium with "ordinary mind and body speech" (an image of material plutonium being made void by the already insubstantial and abstract human speech and thought).

As such, "Plutonian Ode" culminates in an implied process of disarmament fulfilled by its closing lines. The entire structure of the poem is directed at this revelation, accelerating in pace to then slow down and finally become measured with the rhythmic caesurae of the final line. The accretion of line and verse length that is so often significant in Ginsberg's poems here finds its outlet in the structure of that final line, which slowly elaborates on the verb to go: "gone out" becomes "gone beyond" then

---

57. Ekaku, "Text of the Heart Sutra."

"gone beyond me," narrating a gentle diffusion of the human breath as well as reflecting back on the placidity of that chanting in the lack of fricatives. Self-reflexively describing its own progress as well as making the phonetics of its diction parallel the process it narrates, "Plutonian Ode" places poetry at the heart of peaceful protest. Cumulatively, the poems Ginsberg composed while protesting the atomic industry are leveled at defusing the destructive potential of nuclear technology. By repetitively emphasizing the human experience and by positioning the wonder that attends the sublime in the immanent power of the elements involved in nuclear fission, Ginsberg suggests that the transcendent experience provided by regarding nuclear technology need not be inextricable from an understanding of how that technology can be abused.

By yoking the Space Race to the atomic bomb, Ginsberg's writing establishes a dialogue between exploration into, and destruction from, outer space and underscores how technology is used by humankind, rather than the aesthetic effect that same technology might have upon its spectator. Rather than meeting the sublime scene of the atomic cloud with a rapture that turns inward as in Laurence's account, Ginsberg's poetry both locates wonder within the human body in poems such as "POEM Rocket" while equally considering the massively detrimental effects of nuclear technology in work such as *Plutonian Ode*. His work invokes the sublime experience as it is so often associated with the overwhelming scale of vertical spectacle, only to then attach the wonder housed by the sublime to poetic form and the human body. In this sense, Ginsberg's work rejects the Kantian idea of a masculine reordering of the world that Christine Battersby describes as a response to the "sensory, conceptual and imaginative conflict" generated by the sublime, a process that works through "comparison (or judgement)" giving way to a "'supersensible' order that does not have the human 'I' at its source."[58] Battersby notes that, in the face of the sublime, "The elite male re-orders the chaos" that has been created by "displacing" individual primacy, "containing the apparently uncontainable," and thus assimilating the sublime moment.[59] This description of the Kantian sublime is at odds with Ginsberg's poetry, in which it is precisely the continued existence of bodily experience (even when that bodily experience is granted as impermanent by Buddhist thought) that is maintained through the sublime moment.

The parity between the Buddhist Sutra and poetic form as vocalizations of spiritual intent facilitates Ginsberg's underlying suggestion that

---

58. Battersby, *Sublime*, 33–34.
59. Ibid., 34.

nuclear technology need not abet the mythology of a regenerative violence. Rather, the poet's later work makes meditative chanting the primary mode of a neutralization, through which the destructive role played by nuclear technology in American ways of life is challenged. Ginsberg's poetry and prose that takes nuclear technology as its topic extends Sinclair's response to scientific and technological advances, and his vertical imagery specifically reacts to a chasm in existing ideas of what is possible. While Ginsberg's writing magnifies Sinclair's questions of the desirability surrounding new technology, unlike *Oil!*, poems such as those collected in *Plutonian Ode* employ vertical imagery to unravel the urges that dictate the abuse of that same technology. Ginsberg's work marks an emerging desire to critique a specific confluence of national events and, at the middle of this book, reveals a shift from vertical imagery as a means of registering uncertainty to vertical imagery as an expression of protest.

While the atomic bomb shattered the idea of the world as "always already there," the splitting of the atom undermined not only previous scientific knowledge but also destabilized the entire epistemological framework of Edmund Husserl's consciousness based upon a "life-world" that continues indefinitely.[60] Ginsberg's writing reacts to this epistemological damage, but it also unravels the vertical spectacle as the site of knowledge based not upon how to destroy but how to rehabilitate transcendent experience. Verticality becomes, then, not merely a sense of waiting for the sky to fall, but a sense of anticipating what the implications of that kind of disaster might be. Rather than simply waiting for bombs to descend from the sky or detonate in spectacular vertical visions, Ginsberg's work engages with the concept of the sublime to suggest that the skewed temporality of a seismic vertical event might be expressly creative rather than destructive.

---

60. Nye, *American Technological Sublime*, 228.

CHAPTER 5

TRAVERSING VERTICAL SPACE

Philippe Petit's Wire-Walk, Danger, and Transformation

REPRESENTATIONS OF PHILIPPE PETIT'S WIRE-WALK

In James Marsh's *Man on Wire* (2008), Philippe Petit's wire-walk of 7 August 1974 is dramatized both as a shadow and an inverted version of 11 September 2001. In one of its most brazen moments, the film visually reverses images of the collapsing towers from 2001 by focusing on the construction of their skeletal structure in the 1970s. The footage conjures the collapse of the buildings by alluding to the Ground Zero photography in which the same vertical

---

Parts of this chapter originally appeared as the 2011 article "'Going Backwards in Time to Talk about the Present': *Man on Wire* and Verticality after 9/11," *Comparative American Studies*, 9:1 (March 2011), 3–20. Since the publication of my article, there have been several examinations of *Man on Wire* that pursue similar lines of inquiry, including Chris Vanderwees's "A Tightrope at the Twin Towers," Hamilton Carroll's "September 11 as Heist," Martin Randall's "'A Certain Blurring,'" and Emily Murphy's "Life on the Wire." Sinéad Moynihan has also written about some overlapping themes, especially in Colum McCann's *Let the Great World Spin*, in her "'Upground and Belowground Topographies.'" At the time of writing this book, Robert Zemeckis's film dramatizing Petit's Walk, *The Walk* (2015), was yet to be released.

steel girders are shown sticking out of the smoking wreckage of the Twin Towers. Ground Zero photography—captured most famously by Joel Meyerowitz, who was granted almost exclusive access to the site—has become the most prevalent visual shorthand for the 2001 attacks by figuring what Art Spiegelman refers to as the "glowing bones" of the World Trade Center emerging from rubble.[1] Most notable here is Meyerowitz's "Assembled Panorama of the Site from the World Trade Center Looking East," which consciously frames a devastating scene of wreckage in order to effect a haunting beauty, chiaroscuro emphasizing the jagged girders as they poke through ethereal smoke. *Man on Wire*'s 1970s footage is both the opposite of and a palimpsest to Meyerowitz's collection of photographs: despite one depicting creation and the other ruin, the two are uncannily similar. Taking this Ground Zero photography as its reference point, *Man on Wire*'s inclusion of various still images and original footage portraying the construction of the Twin Towers forges a crossover between aesthetic pleasure and a site of destruction, making the events of 1974 into an oblique address to those of 2001.

The text from which *Man on Wire* was adapted—Petit's memoir, *To Reach the Clouds: My High Wire Walk between the Twin Towers*—was published only a year after 9/11, and after 2001, Petit's performance has become the topic of a variety of other literary and visual texts. This chapter considers the ways Petit's performance has been represented after 9/11, suggesting that texts invoking the wire-walk simultaneously betray how the events of 11 September 2001 forcibly pushed verticality into consciousness via the sight of the planes flying into the World Trade Center and the distressing reality of the bodies that fell from the towers. I argue that these texts re-mediate Petit's performance of 1974 precisely to summon the experience of watching a wire-walk firsthand as, in Paul Auster's terms, a "lure" away from thoughts of danger and death. Auster's description of wire-walking asserts that the "good" wire-walker "strives to make his audience forget the dangers, to lure it away from thoughts of death by the beauty of what he does on the wire itself." Further, and tellingly, high wire-walking "is not an art of death, but an art of life—and life lived to the very extreme of life."[2] The schema Auster describes is instructive. *Man on Wire* attempts to dislodge the allusions to 9/11's deathliness embedded in its early images of the World Trade Center's skeleton with repeated affirmations of Petit's beautiful creativity throughout

---

1. Spiegelman, "Sky is Falling!," ii.
2. Auster, "On the High Wire," 253.

its narrative. Depictions of Petit by those such as Petit's girlfriend of the time, Annie Allix, as "so excessive, so creative" and the suggestion that "each day is like a work of art for him" are annexed to assertions that variously declare how his wire-walk was "so, so beautiful."[3]

Man on Wire moves between representations of Petit's wire-walks (although, as I discuss later, it does not offer original video footage of the Twin Towers wire-walk itself) and those of the events leading up to 7 August 1974.[4] The film's reconstruction of that day begins in the early hours of the morning, when Petit and a group of his associates broke into the World Trade Center, proceeding in two parties past the unfinished floors to the roofs of the North and South Towers. From the North Tower, Jean-Louis Blondeau used a bow and arrow to shoot a cable across to the South Tower. After spending several hours securing the wire, Petit stepped out onto the cable and, in his own words in the film, "danced at the top of the world" for over forty-five minutes. Verticality in its post-9/11 context is the foundation for *Man on Wire* not merely via the film's early images of the skyscraper being constructed but in the very structure of its narrative. The plot follows Petit and his accomplices as they enter the World Trade Center's basement and climb up to the roof, conducting an ascent through the skyscraper that reverses the way in which the same buildings were attacked in 2001 from above. The film's depiction of Petit walking between the towers at 1,350 feet also alludes, without fully succumbing, to the memory of bodies falling from the towers on 11 September 2001. Finally, the film engages the same visual dynamic as that operating during the events of 2001; during both Petit's walk in 1974 and the later terror attacks, witnesses were forced to angle their line of sight upward to see what was happening above them. Throughout *Man on Wire,* the events of 2001 are summoned at critical moments of the plot and then displaced by those of 1974 to offer Petit's walk as a palliative to the myriad traumas of 9/11.

Crucial to the following discussion is the idea that *Man on Wire* participates in a temporal representation directly stimulated by the 2001 attacks. In these terms, the film's opening images of the World Trade Center under construction articulate a desire for the Twin Towers not to have collapsed

---

3. Subsequent quotations from interviews in *Man on Wire* are referred to with the speaker in question named in the main text.

4. See also Mackay and Mitchell, "Re-constructing 'Le Coup,'" for an analysis of the film's form in relation to Derrida's concept of "the event," ideas of exceptionality, and time.

that is common in post-9/11 narratives.[5] After its unveiling, the World Trade Center was variously regarded as either an architectural monstrosity or a symbol of the banal face of corporate America. Paradoxically, the buildings would come to be appreciated, and even eulogized, after their destruction.[6] In this sense, *Man on Wire* shares with other post-9/11 films an emphasis on the World Trade Center in order to engage the "profound emotional need" of a great many Americans to have their "incipient nostalgia" fulfilled by images of the Twin Towers.[7] This desire couples with a long-standing preoccupation with the extreme reaches of the American skyscraper, and partly accounts for the retrospective fascination in literary and visual arenas with the absent towers. This preoccupation is by no means limited to novels and films; after 2001 the Twin Towers suffused the public domain in the form of commemorative items such as the souvenirs Marita Sturken discusses.[8] But the significant cultural cachet accrued by *Man on Wire* and other texts that represent Petit's walk is not fully explained by mere wish fulfillment. Rather, the film's success rests upon the way in which it uses nostalgic sentiment not only to recall the Twin Towers intact, before they fell, but also to suggest the inverse: their construction. In a gesture of temporal disruption that this book has argued is frequently attached to vertical imagery, the shots of the towers' creation in the early 1970s pursue a temporal reversal in which the same steel girders are erected into towers rather than collapsing into the Ground Zero site.[9]

Man on Wire, while participating in the nostalgia discussed above, also enters into dialogue with various competing demands placed upon the way 9/11 is conceived: as exceptional and unprecedented, or predictable and foreseen, or somehow all of these at once. The 2001 attacks were quickly embedded in a popular rhetoric that attempted to fix the events of that day as unparalleled while describing the attacks in terms of a destabilized sense of New York as a "safe" place with a permanent and secure geography.[10] This very idea—of the city as somehow materi-

---

5. Schneider, "Architectural Nostalgia"; Negra, "Structural Integrity."
6. Simpson, *9/11*, 58–59; Sturken, *Tourists of History*, 201–2; 222.
7. Schneider, "Architectural Nostalgia," 29–41.
8. Sturken, *Tourists of History*, 1–4; 26; 212–15; 217. Further considerations of absence, memorialization and commemoration, and trauma include Simpson, *9/11*; Miller, "'Portraits of Grief'"; Kaplan, *Trauma Culture*; Savage, "Trauma"; Sherman, "Naming and the Violence of Place."
9. Murphy discusses the temporality of representations of Petit's walk, including the role of nostalgia; see her "Life on the Wire," 68–73.
10. For discussions of this rhetoric, see Greenberg, *Trauma at Home*; Sherman and Nardin, *Terror, Culture, Politics*; Butler, *Precarious Life*; Simpson, *9/11*. Moynihan

ally stable—is itself a short-term perspective occluding New York's historical cycles of destruction and rebuilding, a process named by Max Page as the "creative destruction of Manhattan" that is refracted as early as 1907 by Henry James's description of the city's new skyscrapers as "provisional."[11] It also ignores the prophesied disaster that has been embodied in various ways from the earliest days of the new skyline, in ideas such as the skyscraper's "architectural death-wish" and in the rich heritage of texts that have imagined New York's destruction, such as early science fiction narratives depicting "a landscape of sublime ruins."[12] The idea of New York as somehow immutable and immune to external threat was also volubly contested by theorists such as Jean Baudrillard and Slavoj Žižek, who famously intoned, "We have dreamt of this event," and "For us, corrupted by Hollywood, the landscape and the shots of collapsing towers could not but be reminiscent of the most breathtaking scenes in big catastrophe productions."[13] Like these discourses that inscribe 9/11 as both singular and/or somehow inevitable, *Man on Wire*'s aural narrative insists Petit's walk was both "impossible" and also his "destiny," emphasizing a particular determinism via assertions such as Allix's statement, "It was as if [the towers] had been built especially for him."

While its representation of the wire-walk mirrors the terms of describing 9/11, *Man on Wire* also works against a post-9/11 language of catastrophe by emphasizing Petit's performance in stark contrast to the destruction of 2001. In this sense, the film recapitulates David Chelsea's 2002 double-spread graphic narrative "He Walks on Air," which was published as part of a graphic collection expressly setting out to confront 9/11. "He Walks on Air" is a light-hearted depiction of the walk in which the graphic form is harnessed to defuse the tensions surrounding verticality after 9/11, framing Petit's walk in a simple and visually striking narrative that would not be out of place in a children's story.[14] Like *Man on Wire,* David Chelsea's text both draws attention to, and attempts

---

examines the idea of 9/11 as "a moment of rupture" in relation to McCann's *Let the Great World Spin* ("'Upground and Belowground Topographies,'" 270).

11. Page, *Creative Destruction of Manhattan*, 1–3; James, *American Scene*, 77.

12. Van Leeuwen, *Skyward Trend of Thought*, 54; Yablon, "Metropolitan Life in Ruins," 309–10. See also Page, *City's End.*

13. Baudrillard, *Spirit of Terrorism*, 5; Žižek, *Welcome to the Desert of the Real!*, 15.

14. Mordicai Gerstein's children's story *The Man Who Walked between the Towers* shares the awed and slightly credulous tone of Chelsea's narrative and also attempts to dispel the anxieties generated by 9/11, but in a way explicitly directed at the children living in New York.

to flatten out, the parallels between 9/11 and Petit's walk. The opening panel affirms the date as "August 7 1974," but the narrative also alludes to 11 September 2001 in its description of "the calm of an ordinary morning" that is "shattered" by "a new and startling sight." Petit's act is textually signaled as a "dance," a "walk for happiness," while the narrative's visual cues allude to the terrorist attacks some thirty years later. Petit's performance is made a feat in direct juxtaposition to 9/11 that flirts with "certain death" but results only in a "display" of "art"; the reality of "certain death" for those trapped in the Twin Towers remains silent, but tangible, in Chelsea's emphasis on the parallels between the two events. The panel's speech bubbles, arising from the base of the Twin Towers, declare: "Look! Up in the sky!" and "It's a plane!," blurring the disparity between the events of 1974 and 2001 further through the common experience of witnesses being forced to crane their necks upward to see what was happening.[15] "He Walks on Air" uses Petit's walk to transpose a "display" of "happiness" onto the memory of the terrorist attacks; like Chelsea's text, *Man on Wire* proposes ascent as a performance with the power to incite euphoria in those who watch. In the film, seeing the wire-walk makes the witness "happy, so happy" (Allix's words), and Petit recalls that, at the top of the tower, "I suddenly had hope and joy." Both Chelsea's narrative and *Man on Wire,* which consistently emphasizes the wire-walk as a "beautiful show" (Allix) at odds with the collapse of the Twin Towers, suggest that the upper limits of the World Trade Center are the stage not for destruction but for a creative act.

In Chelsea's narrative, Petit's performance harmoniously unifies those who watch, in direct contrast to the disturbing collective experience of witnessing planes fly into the World Trade Center. *Man on Wire* turns Chelsea's emphasis on a shared experience into part of its very form, representing the wire-walk via a spectrum of media and the collaboration of diverse perspectives and voices. This collage form incorporates not only interviews with Petit and each of his accomplices but also those law enforcement officers tasked with terminating the wire-walk. Various anonymous witnesses to the event are also represented, albeit seen rather than heard, by the inclusion of photographs taken from the ground by pedestrians as Petit performed. The film's diverse formats also move between different times, the two most distinct of which are the 1970s and the post-9/11 "present day" (or rather, 2007, when the film was shot). *Man on Wire* separates these forms from one another through the strategy

---

15. Chelsea, "He Walks on Air," 10–11.

of presenting its reenacted sequences in black and white. Petit's walk is the uniting thread through all these disparate viewpoints and moments, which share a common desire to interpret the events of August 1974 and decipher the "alien" figure of Petit who erupts onto the urban landscape.

Cumulatively, the film's complex formal qualities work to conflate height with a desirable space. While creating a tissue of different temporalities and perspectives, the film's explicit aural narration links a celebratory tone to the theme of verticality in its plot through a score that alludes to ascent or height. Ralph Vaughan Williams's "The Lark Ascending" (a piece that was also, tellingly, played during the ceremonial opening of the New York 9/11 Memorial in September 2011) accompanies the shots of Philippe climbing trees as a child, and Edvard Grieg's "In the Hall of the Mountain King" plays while Petit and one of his accomplices "escape" to the roof from the policeman guarding the top floor of the South Tower. Reaching the apex of the towers is equated with an act of colonization via the subtle allusion provided by the musical strains of "In the Hall of the Mountain King" and confirmed by Allix's suggestion that Petit seeks to "conquer" the World Trade Center. Although oblique, the allusions to a bird "ascending" and the idea of the "mountain king" in these titles are brought to the fore by their timing: played at key moments in the film, they suggest that the narrative's stress falls on Petit's act as a kind of flight that allows supremacy over a vertical structure.

DEFLECTING THE TERRORS OF VERTICALITY: VERTICAL SPACE, FALLING BODIES, AND "HARMLESS CRIME"

If *Man on Wire*'s conflation of vertical space with positivity is immediately apparent, the film goes even further to suggest a vertical frontier exempt from, first, common law and, second, the memory of falling bodies after 9/11. For Petit, the skyscrapers' vertical reach offers the allure of "taking certain liberties" (Allix's words). The film represents movement upward through the World Trade Center both as a literal escape as Petit ascends ahead of the security guards who seek to eject him from the building, and as a metaphorical rising above conventional modes of authority in the suggestion that the wire-walk is exempt from a normal paradigm of legal process. Throughout the film, Petit and those close to him insist upon the wire-walk as both illegal and permissible: Allix calls the walk an "adventure" and also suggests that "what excited" Petit "most" was that "it was like a bank robbery." By making Petit's wire-walk a foil to the events of 2001, and by using verticality not only to inform the wire-walk itself but also the other questions of transcendence, illegality, and terror that underpin the film, *Man on Wire* reinscribes verticality after 9/11 as distinct from falling bodies and falling towers.

The film recasts the towers as a site for a joyous performance, rather than a target for terrorism, by using a frame of reference—the heist or caper film—that allows it to summon the same themes of illegality inherent to 9/11 while ultimately characterizing Petit's act as a defensible crime. An expansive genre reaching back to *Alias Jimmy Valentine* (1915), the heist or caper narrative is the template for variations upon the same theme in the noir cinema of the 1940s and 1950s such as *Asphalt Jungle* (1950), *The Killing* (1956), and *Bob le Flambeur* (1956), later light-hearted robbery films such as *Ocean's Eleven* (1960), *The Thomas Crown Affair* (1968), and *The Italian Job* (1969), and more recent contributions such as *Heat* (1995), *Out of Sight* (1998), and *Bandits* (2001). Most relevant to *Man on Wire* is the comedy caper of the 1960s, which narrates "white-collar crime or theft by characters who are not thieves." This genre "glamorizes crime as stylish adventure, complete with fashion, romance, and sex" to provide a "vicarious escape" from convention.[16] The popularity in the 2000s of remaking such "classic" heist films such as *The Italian Job* and *Ocean's Eleven* suggests that, beyond the case of *Man on Wire*, the depiction of "harmless" crime is a more general foil to the issues circulating in the twenty-first century: a "vicarious escape" from the realities of a world in which terrorism has become the default mode of international attack. *Man on Wire*'s overtures to the heist genre make Petit's act synonymous with the kind of criminal activity within films in which theft is to some extent narratively justified by the inherent likeability of its characters and the juxtaposition of theft to more extreme crimes. The way that theft is glamorized by the heist film and, at the same time, understood as less reprehensible than crimes such as murder and rape (which tend to be viewed in ethically polarized terms) is nonetheless problematic and, as I discuss in detail later, anticipates the uncomfortable crossover between art and terror in *Man on Wire*.

Significantly, the appropriation of the heist template in *Man on Wire*'s opening shots gives way to myriad other filmic inflections such as the "talking head" documentary format and the "city symphony" form of the 1920s discussed later in this chapter. By swerving between formal and generic conventions, *Man on Wire* incorporates a range of discourses

---

16. Von Dassanowsky, "Caper of One's Own," 108.

that all underline Petit's walk as both similar to, and yet radically different from, the crimes committed in 2001. This process is negotiated by the film's opening, which, by marrying recognizable elements of the heist movie to forms of conspiracy, generates the expectation that a terrorist plot will be revealed in the course of its narrative. In the sequence, close-ups focus on a calendar marked "LE COUP" on 7 August 1974; the camera pans down from the calendar to a table covered in blueprints for the World Trade Center; a man slots an arrow into a narrow tube as a television screen next to him portrays the unmistakable façade of the White House and then cuts to footage of a press interview with President Nixon; and figures load a box into a van. The close up of *coup* knowingly plays upon the several meanings it has accumulated in English: as well as its primary meaning of a strike or blow, *coup* also refers to an overthrow or fall, as well as, most recognizably, a shorthand for *coup d'état*, a seismic shift in government executed violently or illegally. The sequence is especially linked to this final, and political, meaning; the close-up on "LE COUP" precedes the images of the arrow and two metonymies of the American political process (the White House and the President) in order to foreshadow a violent struggle and political overthrow that never comes.

The visual emphasis of "LE COUP" written on the calendar signals the central theme of illegality while revealing how the film tries to dispel anxieties left by the 2001 attacks. Proleptically indicating an attack on the World Trade Center or Presidential assassination that will not actually be fulfilled, the film's opening allusion to terrorism as the foundation for the illegal vertical ascent depicted within its main narrative is just that—an allusion, which later dissolves as the film insists on Petit's walk as a "harmless" event. Throughout its narrative, the film avows its protagonist is a man who is "so excessive, so creative" that "each day is like a work of art for him" (Allix's description); the film calls to mind the illegal action of terrorists but supplants the towers' destruction with Petit's transgressive but celebratory performance in order to destabilize a relationship between terrorist action and verticality. The marketing for *Man on Wire* worked similarly, replacing the specter of terrorism with art in its tagline of "1974. 1,350 feet up. The artistic crime of the century." This phrasing carefully establishes the film's location as "1,350 feet up," connecting extreme height with both a creative act (artistry) and illegality (crime). At the same time, "1,350 feet" acts euphemistically, avoiding explicit references to "the World Trade Center" or "the Twin Towers," and the figures "1974" and "1,350" both echo the unspoken numerical notation "9/11." The tagline simultaneously invokes 9/11 and jettisons it from the film's frame of reference, suggesting that 9/11 is both central to the narrative and also something it wishes to move past. Once more, the crossover between creativity and destruction effected by the construction and post-9/11 photos of the Twin Towers comes to the fore.

The invocation of terrorism as a bluff in *Man on Wire*'s opening sequence, which works to dispel anxiety by producing a foil to the specific terrorism of 2001, is condensed by its overt reference to "Americanness." Jean-François Heckel recalls that "to look like an American, I had lots of pens in my pocket": simulating Americanness is indicated as the only way to infiltrate the building when one bears criminal intent. While here the film seems to indulge a conflation of terrorism and foreignness, in fact, it immediately defuses the insidiousness of this coupling via the light-heartedness and prosaicness of bearing "lots of pens" as the shorthand for being an American.[17] The final shot of the opening sequence depicting the group loading the back of a van with a box works in a similar way. The shot of the van introduces the sequence following the group traveling through the streets of New York prior to their ascent of the building from the basement of the World Trade Center, the underground space where Ramzi Yousef and Eyad Ismoil detonated a bomb on 26 February 1993. Meanwhile Petit's voice-over obliquely equates vertical motion with a kind of monstrosity as he recalls feeling that "the horizontality of driving through the streets of Manhattan suddenly became slant" and describes the feeling of being "engulfed by the monster" when they approach the building. The film's visual emphasis on the truck and its verbal references to the basement refer back to the earlier act of terrorism at the World Trade Center nearly ten years prior to the 2001 attacks. Although "the monster" is ostensibly Petit's compulsion to complete the wire-walk, the term "monstrosity" is also linked to the subterranean space where the 1993 attacks occurred. However, the group's descent into the bowels of the World Trade Center is not made congruous with the 1993 terrorism; by attempting to alleviate the tensions summoned by references to conspiracy after 9/11, *Man on Wire* actively seeks to dispel the specter of the 1993 attacks from its narrative.

The subterranean garage is made into the access point for Petit's "harmless" act, and allusions to the 1993 terrorists' entrance from below

---

17. The conflation between terrorism and foreignness exists most recognizably in popular American texts like the television series *24* (2001–14); thrillers such as *Patriot Games* (1992), *The Siege* (1998), and *Body of Lies* (2008); or action films like *Die Hard* (1988), *True Lies* (1994), and *Air Force One* (1997). It also pervades the post-9/11 development of national rhetoric; see Kaplan, "Homeland Insecurities," 87.

the building are further refigured by the comedic depiction of Petit's second visit to the World Trade Center, "spying on the Twin Towers." As Walter Murphy's playful and up-tempo "A Fifth of Beethoven" plays, Petit enters the building from the basement, looking at "what commercial vehicles go underneath for freight delivery" while trying to "disguise" himself. When doing so, he injures his foot—a shot comically zooms in on the offending nail—and finds that, on crutches, he is now able to enter the building without rousing suspicion. The formula of a heist film, as in the beginning, summons the phobias and tensions resulting from the 9/11 attacks, but this sequence reframes the echoes of the 1993 attack by replacing it with Petit's emphatically non-ominous "spying"; his entrance is from below the building but is a radical departure from the terrorists' infiltration of the World Trade Center. In addition to the comic sequences of the film, Petit's performance throughout in interviews and reenactments is charismatic, offering a humorous physicality. And not merely in this sequence, but through the entire film, the process of moving up through the building is a ludic and comically elaborate escape from the hand of the law, lightening the film's narrative parallels with crime. As Petit describes hiding on an upper floor of the World Trade Center, awaiting the morning of his wire-walk when a policeman enters the space and conducts a survey of the area, the film reconstructs a playful game of "hide-and-seek" in which Petit hides behind a pillar and, as the policeman rounds the corner, avoids the other man by moving in the opposite direction. The whimsical tone continues as Petit and Blondeau, depicted in silhouette carrying equipment up a staircase to the "heist" theme music, walk in a ponderous and humorously overemphasized way. Rather than equating the wire-walk with the terroristic attacks of 1993 from below the building, or those of 2001 from the sky, the formal qualities of the film and its comic tone work to counter a memory of these attacks.

Man on Wire is equally intent to dispel anxieties left by 9/11 in its depiction of Petit's walk, in stark contrast to the distressing reality of bodies falling from the Twin Towers nearly thirty years later.[18] And, while the specific visual portrayal of figures falling in recent texts cannot exist in a vacuum free from the photographs of falling people on 11 September 2001, the spatial metaphor of descent reverberates through work that attempts to understand trauma. Eleanor Kaufman argues that the conspicuous link between trauma and images of falling exposed by postwar

---

18. See also Frost, "Still Life"; Lurie, "Falling Persons and National Embodiment"; Kroes, "Ascent of the Falling Man."

literature is explained by "a new and more aerial form of spatial perception" guaranteed by the always imminent threat of something falling from the sky.[19] But falling also lays claim to the very texts that attempt to understand wider and abstract concepts of psychic trauma; Cathy Caruth uses the metaphor of a fall for the traumatic moment, to indicate both the "abysslike structure of trauma" and how "a crucial and potentially debilitating undecidability is set into motion."[20] The fall—a moment that might or might not end in fatality, might be terminated by a parachute or end in catastrophic impact—concretizes Caruth's idea of trauma as that which exceeds the parameters of foresight and comprehension.

In post-9/11 narrative, the recurrence of falling imagined "safely" exposes a specific longing for the events of 2001 to be undone. Petit's almost magically suspended body offers a way of reimagining falling that is sought by earlier post-9/11 novels, in narratives that attempt to halt, reverse, or replace the deadly descent from the Twin Towers. Notable among these is the text with which I began this book, *Extremely Loud and Incredibly Close,* which strives to hold static the falling towers and the falling people within its narrative. Petit's performance also reverberates, obliquely but tellingly, through Don DeLillo's *Falling Man* (2007), in which the titular performance artist enacts staged falls from skyscrapers while attached to a harness, supplanting falling with performance art.[21] In a similar vein to Foer's magically suspended body or DeLillo's figure that falls incompletely, *Man on Wire* negotiates the presence of falling recalled by the location of the Twin Towers by foregrounding Petit's elevated but not descending body. The problems encountered by this emphasis on Petit suspended in space are dramatized by "Soaring Spirit," the artwork by John Mavroudis and Owen Smith that comprised the front and inside covers of *The New Yorker* on 11 September 2006. Here, Petit's wire-walk is represented as an exceptional event that does not merely frustrate falling but might even overcome death; "Soaring Spirit" portrays a kind of impossible suspension and clearly tries to use the wire-walk in order to commemorate, and move past, 9/11.[22] The front cover depicts a wire-walker over a white background; the first inside page redraws the same figure, now over a background showing the Ground Zero site. Both the context of the cover, which was created for an edition published on

---

19. Kaufman, "Falling from the Sky," 44.
20. Caruth, *Unclaimed Experience,* 5–7; 73–90. Kaufman, "Falling from the Sky," 50; 47. See also Mauro, "Languishing of the Falling Man," 589.
21. See also Rowe, "Global Horizons in *Falling Man,*" 131.
22. See also Frost, "Still Life," 196–98.

the fifth anniversary of 9/11, and the overt signposting of its title indicate that Petit's performance is here retroactively deployed as a commemoration to the New York population.

The wire-walk is used to obliquely address the horrific events of 2001, the image's title both explicitly referring to Petit's act as "soaring" and metaphorically suggesting that an American "spirit" might overcome the memory of that day. The description of Petit's walk as a kind of ascension aligns with the yearning for unencumbered elevation in DeLillo's and Foer's narratives, yet falling continues to haunt the image's various modes of representing absence. In the first image, the man is suspended over a white page; absence here is approximated by a complete lack of color and no visual context. In the second, the figure is ostensibly provided a context, suspended over a background, but this background only shows another absence via the gaping hole where the Twin Towers once stood. Despite the effort to displace the memory of 9/11 with an act that in one sense provides a visual opposite to falling bodies, the doubling of an identical male figure and the recurrent representations of absence make "Soaring Spirit" notably circular, and effect a repetitive return to the disturbing idea of absence. In the first image, this circularity works against the expressed intent of the cover to surmount—to "soar" above—the memory of 9/11 and, in fact, only makes the figure appear more precarious, more threatened; in the second image, absence only returns the viewer to the inescapable physical lack of the Twin Towers. While the underlying motivation for anachronistically summoning the events of 1974 is common in post-9/11 narratives such as "Soaring Spirit," *Man on Wire* attempts to subvert the problematic association of absence and falling with the terrorist attacks by representing Petit's walk as an explicitly "magical" act.

This strategy is embedded in the film's formal construction, which renders the walk extraordinary not only in its verbal narration of an "impossible" and "magical" act, but compounds its verbal references to magic through its formal syntax. Fusing the form of films like *United 93* and *9/11*, *Man on Wire* incorporates original footage, reenactment by those involved in the wire-walk, and sequences using professional actors.[23] The film's spine of "real" footage comprises "talking head" interviews with those involved in Petit's wire-walk, videos of Petit conducting wire-walks across the Notre Dame Cathedral in Paris and the Sydney Harbour

---

23. For discussions of the development of documentary as a genre fusing different formats, see Austin and de Jong, *Rethinking Documentary*, especially 2ff.; Bruzzi, *New Documentary*, especially 1–8.

Bridge as well as planning the World Trade Center "heist," still images of Philippe and his team at the Twin Towers researching and then conducting the wire-walk, photographs of Philippe as a child, images and video footage of the World Trade Center being erected, an interview with the policeman who "captured" Philippe, and original media footage from a helicopter over the Twin Towers as well as audio clips from radio reports of Petit's walk. Standing between recognized modes of "documentary" and "drama," the film at once documents the history of the World Trade Center and also prioritizes a kind of mythic fairytale concerned with what Petit repeatedly calls an "impossible" act. The original photographs used to represent Petit's walk and the acted or reenacted sequences instigate the performative implication that eventually the 1974 walk will be shown via original video footage. Although the film provides computer generated images approximating Petit's viewpoint from the top of the World Trade Center, the implicitly promised original video footage never comes. Crucially, the film's collage-form and its syntax both work to obscure this fact; *Man on Wire*'s performative sleight of hand works because it is a hybrid text spanning several different genres as well as being formally complex, and the film's fusion of formats that move between different temporal moments makes it easy to miss the fact that original video footage of the Twin Towers walk is missing. As such, the film's form itself is a performance as much as its narrative is about Petit's particular performance in 1974, and these representational energies converge in the film's effort to distract from the distressing memory of death by falling from a great height.

Both narrating Petit's act and also obliquely addressing the events of 11 September, *Man on Wire* stands in distinction to other post-9/11 films such as *9/11* (directed by Jules Naudet, Gideon Naudet, and James Hanlon, 2001) and *United 93* (directed by Paul Greengrass, 2006), which explicitly document the events of 2001. *9/11* maintains a sense of verisimilitude throughout by only utilizing "original" footage from the day or interviews with those involved; other documentaries favor forms of reenactment, such as Bruce Goodison's *Flight 93: The Flight That Fought Back* (2005) and Chris Oxley's *Let's Roll: The Story of Flight 93* (2002), both of which depict the failed hijacking of United Airlines Flight 93. At the extreme of this form lies *United 93*, which uses a full cast of actors to play the passengers and terrorists on board the flight in a real-time "speculative" narrative.[24] Films that directly represent the events of 2001,

---

24. Ward, "Drama-Documentary," 191–92.

such as *United 93* and Oliver Stone's *World Trade Center* (2006), make efforts to capture the resilience and heroism of those caught in the fallout from the attacks. Films that nonfictionally deal with post-9/11 society often deal in a "rhetoric of panic" to articulate anxieties surrounding security and terrorism.[25] Both these types of film work to secure a cathartic effect for their audience by emphasizing American heroism in the face of the most extreme crimes. *Man on Wire* shares this common desire for catharsis but is unique for its choice of subject. By employing verticality as its cardinal theme, the film rouses the disturbing memories of 9/11 but tries to move beyond a "rhetoric of panic" by transcribing that same verticality as a less complicated source of pleasure.

PETIT'S PERFORMANCE AND THE "EDGE OF LIFE"

While in many ways *Man on Wire* works to deflect the anxieties surrounding falling bodies and terrorist plots after 9/11, its entanglement in the theoretical crossover between terror and art creates a problematic complicity in which the events of 1974 and 2001 to some extent bleed into one another. In its formal and thematic qualities, and its aural narration, the film modulates the wire-walk as a creative outpouring and a consummate performance in such a way that Petit's act becomes enmeshed in issues of transgression and danger. The following discussion examines the ways in which the film negotiates these problems by attempting to shift the very ground on which artistic endeavor and terrorist acts are defined.

Transgressive acts are typically understood to imply "a hostile and intentionally destructive reaction" against "boundaries."[26] *Man on Wire* attempts to circumvent this association by making Petit's act into less of a "destructive" transgression and more of an attempt to undermine what is construed as permissible behavior. Anthony Julius suggests that transgressive art expresses a common artistic desire to frustrate the law, which is assumed to impose "constraints on the imagination" and therefore have "no place in art."[27] Petit's statements in the film affirm precisely this opinion, and he implies that he should be permitted to complete a wire-walk even if his own death should occur, stating that to fall while performing would only ensure "a beautiful death" in the "exercise of passion." Petit's

---

25. Muntean, "'It Was Just Like a Movie,'" 52; Sánchez-Escalonilla, "Hollywood and the Rhetoric of Panic," 11.
26. MacKendrick, "Embodying Transgression," 140.
27. Julius, *Transgressions*, 8.

words expose the fact that the erasure of "constraints" placed upon non-physical art such as paintings and photography is one matter, but that the assertion "law has no place in art" is far more problematic vis-à-vis performances in which literal bodily danger complicates issues of power.

While Petit's wire-walk effectively ensnared those spectators watching from the ground in the danger of what they witnessed, *Man on Wire* works to absent questions of responsibility from its plot and refuses to voice the fact that had Petit fallen he would have killed not only himself but potentially others watching from the ground. To some extent, the successful outcome of Petit's walk is the most significant reason why the film is able to sidestep these questions. The film nonetheless also allows very little space within its narrative to speculate upon the potential consequences of the walk. Petit's accomplice Jim Moore covers an opaque reference to Petit falling (euphemistically termed "the consequences of anything happening") with a prompt, and more explicit, description of the act, prefaced by the assertion that it was "a really fun adventure to get involved in." Another accomplice, Mark Lewis, notes that the "unforeseen things" worried him more than Petit's capability as a wire-walker, referring to American culture as particularly "litigious." In non-emotive and official language that is particularly distancing, Lewis suggests that the threat was not so much the death of Petit or spectators, but more that of being made "liable" for "involuntary manslaughter" or "assisted suicide." These comments reveal that those involved in staging Petit's walk were aware of the myriad consequences had he fallen, but the comparatively brief time allotted to such questions within *Man on Wire* means that the film distances itself from these consequences.

By foreclosing such questions, and by emphasizing Petit's description of the performance as "a beautiful death," the film actively carves out a place for the wire-walk as exempt from the paradigm of other performative acts. The World Trade Center is made into "a desert island" of "dreams" that fulfills what Petit calls a "need" for "absolute detachment" and "complete freedom." Crucially, the emphasis on the apex of the towers as a "desert island" not only spatially imagines vertical extremity as physically "free" from everyday life but also condenses the idea of the walk as exempt from the legal process and ethical responsibility. As much is echoed by Paul Auster's suggestion that high-wire-walking is somehow beyond a schema of permissive activity since it "is not an art of death, but an art of life—and life lived to the very extreme of life."[28] In terms

---

28. Auster, "On the High Wire," 253.

strikingly similar to Auster's, Petit calls the walk an act at the "edge of life" and suggests that "life should be lived on the edge of life." Petit's strangely tautological sentence, in which the same word—"life"—is both the subject noun and the noun of the adverbial phrase, indicates the similarly problematic transference of vertical space to the "edge" of legal and ethical procedures. Just as Petit's image attempts to make the performance into a state of exception at the fringe of social codes, its own linguistic circularity teeters on the brink of semantic collapse.

Man on Wire's insistence that what Petit did was both a life-affirming act and one that embraced rebellion allows the film to absent questions of responsibility from its plot. But the very avoidance of these questions in *Man on Wire* raises the specter of 9/11, drawing attention to the fact that in many ways Petit's walk summons the terror it is so determined to push away. Comments by the film's director James Marsh, in which he argues that the walk should "stand alone" and in distinction from the 2001 attacks, illustrate the problems of invoking 9/11 while simultaneously trying to repel its inferences. Clearly, *Man on Wire* could not stand free from its audience's awareness of 9/11, and the film's creators would surely have known this. Describing the walk as "beautiful" and "not in any way destructive," Marsh echoes the description of those interviewed within his film by explicitly contrasting Petit's actions with those of the terrorists. He argues:

> What Philippe did was incredibly beautiful. It may have been illegal, but it was not in any way destructive. It would be unfair and wrong to infect his story with any mention, discussion or imagery of the Towers being destroyed. Everyone knows what happened to those buildings. The film has a poignancy for that reason, but not one that needs to be overstated.[29]

Marsh's use of the words "unfair" and "wrong" seems an attempt to camouflage the fact that he must have been acutely aware that 9/11 would at least inform, if not guarantee, the success of *Man on Wire*. More than simply disingenuous, his comments are particularly loaded; using terminology such as "wrong" implies a moral transgression by those who would see 9/11 refracted in the film, despite the fact that *Man on Wire* itself does some of the work of this "infect[ion]" by suggesting that the film's "poignancy" rests upon the events of 2001. Ultimately, the idea that

---

29. Marsh, quoted in Smith, "Wire-Walk Film."

the events of 11 September 2001 could "infect" the film inscribes 9/11 with seemingly viral life; a particularly volatile metaphor, it betrays the underlying desire to at once address and push away the horrific association of 2001.

Marsh, like the film itself, signals the walk's illegality while insisting that it was not "in any way destructive." Crucially, the argument here is apodictic: Marsh can assert that the walk was not destructive only because it is an event safely lodged in the past. Moreover, his comments propose that beauty and destruction are mutually exclusive, whereas the film's depiction of harmless criminality triggers the memory of the 2001 attacks in such a way that art and terror become compatible with one another. Like Marsh, Colum McCann describes his use of Petit's wire-walk in his novel *Let the Great World Spin* (2009) as "a spectacular act of creation" in direct opposition to "the act of evil and destruction of the towers disintegrating."[30] While McCann uncovers what is tacit in Marsh's comments—that Petit's performance is a way of grappling with the later collapse of the towers—he, too, is determined to place the wire-walk and the "towers disintegrating" at opposite ends of a spectrum. And yet, as Frank Lentricchia and Jody McAuliffe indicate, art and terror can coexist in the same performative moment and even share a common drive to "undo" systems of order:

> The desire beneath many romantic literary visions is for a terrifying awakening that would undo the West's economic and cultural order, whose origin was the Industrial Revolution and whose goal is global saturation, the obliteration of difference. It is also the desire, of course, of what is called terrorism. Transgressive artistic desire—which wants to make art whose very originality constitutes a step across and beyond the boundaries of the order in place—is desire not to violate within a regime of culture (libel and pornography laws, for example) but desire to stand somehow outside, so much the better to violate and subvert the regime itself.[31]

Like terrorism, transgressive art (of which performance art is potentially a particularly subversive and visceral format) attempts "to stand somehow outside." While *terrorism* is often used to refer to a specific policy of

---

30. McCann in interview with Johnston, "2009 National Book Award Winner." For a discussion of McCann's novel in terms of Petit's walk, see Cullingford, "American Dreams," 85–86.

31. Lentricchia and McAuliffe, *Crimes of Art + Terror*, 2–3.

systematic terror, it essentially involves terrorizing another sentient being. To some, Petit's walk was an act of "beauty," but it could also have been witnessed with great fear; Allix concedes in the film that the walk was "absolutely terrifying."

Petit's walk and the attacks of 9/11 also share a wider sense of "plotting" in terms of that which is both conspiratorial and artistic. As others have noted, the suicide terrorists who destroyed the Twin Towers in a sense "made" performance art; "there were authors (bin Laden, Atta, etc.)," there was "plot" (a "structure of events with deep narrative inevitability"), there were "thousands of characters," and there was "an audience with no choice but to experience terrorist narrative once that narrative found its true medium of communication."[32] The uncomfortable correlation between terrorism and performance art was made strikingly apparent in the reaction to German composer Karlheinz Stockhausen's description of the terrorist attacks as "the greatest work of art" on 16 September 2001. His statement, which was swiftly retracted, read: "That characters can bring about in one act what we in music cannot dream of, that people practice madly for 10 years, completely, fanatically, for a concert and then die. That is the greatest work of art for the whole cosmos."[33] Unsurprisingly, Stockhausen's comments provoked rapid censure from several quarters; "international action was swift and predictable"; his concerts were cancelled; his daughter declared she would no longer take his name. As Lentricchia and McAuliffe point out, perhaps it was the wistfulness of Stockhausen's comments—the sense that he was transfixed by the event, even envious of the ability to produce a consummate artistic act—that stimulated most outrage.[34] Leaving aside whether or not Stockhausen's comments were intended (or merely received) as offensive, his analogy reveals a porous relationship between art and terrorism that occasionally allows the two to meet.

While I do not mean to imply that *Man on Wire* frames Petit's act as a form of terrorism, or that it seeks a correspondence between the wirewalk and the 2001 attacks, the failure of both Petit's voiceover and also the narrative itself (which insists upon the playfulness and harmlessness

---

32. Ibid., 14.

33. See the *New York Times* account on 19 September 2001, "Attacks Called Great Art"; Battersby, *Sublime,* 21–22.

34. Lentricchia and McAuliffe, *Crimes of Art + Terror,* 6–14. Notably, the authors argue that on "traditional theoretical grounds," "the pain-giving events" of September 11 "are not art," but are careful to highlight how, on these same terms, "images of ground zero in lower Manhattan may indeed deserve to be called art" (9).

of the act) to interrogate the link between performance and ethics means that performativity is granted an exceptionality all its own. Auster implies that wire-walking is exceptional because, unlike other performative art, it is "direct, unmediated, simple, and requires no explanation whatsoever"; its art is "the thing itself, a life in its most naked delineation."[35] What Petit did is regarded both by commentators such as Auster and the creators of *Man on Wire* as a performance that "requires no explanation." Yet by emphasizing verticality as a movement beyond the law, and by eschewing the terrifying possibility of falling in its narrative, *Man on Wire* does not work through the parallels between terror and (anarchic) artistic creativity. The association of falling after 9/11 raises uncomfortable ethical questions about Petit's act; these questions are important, as they uncover his walk operating at a point along a theoretical spectrum of performative acts upon which terrorism also lies.

Rather than address these issues, *Man on Wire* actively invokes the sublime in order to make its representation of the wire-walk less about a fear of impending death and more about a way of processing that same fear into something else. As I have suggested at various points in this book, images of collapse and ruin have plagued the skyscraper consistently throughout its history, ensuring that the tall building is imagined as the source of the sublime in terms of a confrontation with what both seems to exceed spatial limits and also holds the potential for great destruction. The film's insistence on Petit's walk at the top of a skyscraper as profound and transformative calls up long-standing configurations of the Kantian mathematical sublime, discussions of the skyscraper such as Roland Barthes's dissection of the "Sacred" assumed to be innate in the tower, and meditations on a particularly American rapture in a response Nye calls the technological sublime.[36] *Man on Wire*'s aural narrative insists upon Petit's walk as a potent spatial metaphor for transcendence, repeatedly affirming that the act was "magnificent," "magic," and "profound" in interviews with Allix, Blondeau, Heckel, and Moore. Compounding the idea that the wire-walk stands "outside" or beyond the realm of ordinary experience, beyond both the law and the threat of death or falling, *Man on Wire* emphasizes that the walk produces the same transcendent after-effect that secures the sublime experience *as* sublime.

There is a crucial distinction between the wire-walk's impact upon the original firsthand witnesses in 1974 and the effect of *Man on Wire*'s

---

35. Auster, "On the High Wire," 253.
36. Barthes, *Eiffel Tower*, 8; Nye, *American Technological Sublime*, 87–108.

representation on the post-9/11 audience, and the film cannot effect a truly sublime experience (as an overwhelming visual moment in which the beholder is temporarily unable to understand what is occurring before them) for its contemporary viewer. However, when the film positions Petit's walk as a renarration of the events of 11 September 2001, it employs a vocabulary that represents the walk within the parameters of the sublime. Furthermore, soliciting the sublime experience allows the film to depict the wire-walk as exceptional in ways less encumbered by the problems involved in the attempt to render the walk exempt from legal and ethical parameters. In this sense, Petit's walk echoes the similarly vertical and equally awe-inspiring events of 2001, but can invoke the pleasure of the sublime without the moral culpability involved in bearing witness to suffering. I am not suggesting that the issues inherent to the wire-walk (for example, if Petit had fallen and killed others) become irrelevant. Rather, I am arguing that the film's representation of the wire-walk, in which the audience comprehends the performance secondhand and is comforted by the understanding that the event depicted will not end in death, operates as a spectacle that might incite a kind of terror in its audience due to its inherent dangerousness without the beholder necessarily being in physical danger. As I discuss in the final part of this chapter, in this sense the film (as distinct from the original act it depicts) circumvents moral questions concerning the pleasure accrued by the scene.

VISUAL RETROACTIVITY, MODERNIST FILM, AND THE SUBLIME

Man on Wire's representation of Petit's walk as an exceptional moment is not only part of its oblique address to 9/11 but also refers back to a historical conversation between verticality and performance art. On 30 June 1860, the French-born tightrope walker Charles Blondin crossed Niagara Falls on a narrow rope 190 feet above the swollen water. He would subsequently repeat the act on several occasions, accompanied by publicity that trumpeted the event as "Extraordinary" and described how the acrobat would "take a Stereoscopic View Of the Thousands Upon the American Side" when he stood at the middle of the rope between Canada and America.[37] If Harry Houdini's escapology is the most obvious and famous

---

37. Graham, "Blondin, the Hero of Niagara." "Blondin broadsheet," a black-and-white photograph copy of the pamphlet, is held at *Niagara Falls Heritage Foundation* in a collection that includes many of these advertisements as well as photographs of Blondin over the Niagara Falls.

reference point for Petit's translation of circus acts into performance art watched by an unsuspecting public, Blondin is a more accurate ancestor, both in the similarity between the two tightrope performances and also in the way that both Blondin's and Petit's walks were compelling precisely for the verticality upon which they were based. The parity between these walks not only reveals how conquering extreme verticality becomes a resolutely sublime event in its American cultural context, but suggests that *Man on Wire*'s retroactive gestures fit into a long history of sublimity as an exception to normal life.

After 9/11, the portrayal of New York as the site of apocalyptic disaster has replicated itself in a spate of films figuring the city after various catastrophes, suggesting that the real collapse of skyscrapers has been channeled into artistic narratives in order to address the memory of 9/11.[38] Moreover, the idea that acrobatic feats might be used to address anxieties about annihilation and apocalypse has been mirrored in other performances, most notably the show "Traces" by the circus troupe *Les 7 Doigts de la Main* ("The 7 Fingers"). The group describes "Traces" as a fusion of acrobatics, dance, and theater, set in "a make-shift shelter" as "an unknown catastrophe" waits outside and the five occupants live out the ethos that, in "the face of impending disaster," "creation is the only antidote to destruction."[39] Created in "the wake of 9/11," as Lyn Gardner reported for *The Guardian* on 10 March 2010 in an article titled "Can Circus Really Tackle 9/11?," "Traces" is "suffused with a desperate anxiety caused by falling bodies" and is a performance determined by the sense that "there is no certainty tomorrow will come." Balletic and choreographed, yet inspired by the same tradition of death-defying circus acts that Petit's funambulism is, "Traces" affirms the sense of a looming disaster precipitated by the events of 11 September 2001, while also confirming that performance art that exploits the possibility of falling but always holds it at bay is an oblique mode of addressing historical catastrophe.

While others avouch that "the imagined destruction" of New York allows "the urban landscape itself" to "acquire meaning, if not an affirmation of its importance, through its imagined demise," *Man on Wire* is unique for conducting themes of imagined destruction without actually fulfilling them.[40] Colum McCann revealed that thinking about Petit's walk in 1974 made him realize he could "go backwards in time to talk

---

38. The films I refer to include *The Day After Tomorrow* (2004), *War of the Worlds* (2005), *I Am Legend* (2007), *Cloverfield* (2008), and *2012* (2009).

39. Les 7 Doigts de la Main, "Traces."

40. Sturken, *Tourists of History*, 283; Page, *Creative Destruction of Manhattan*.

about the present" whereas, by visually recalling and recoding the tenor of modernist films, *Man on Wire* does not offer the events of 1974 merely as a way to talk about post-9/11 culture but also enters into a certain cultural retroactivity.[41] Posing Petit's wire-walk as a marriage of the two supposedly antithetical modes of art and demolition, *Man on Wire* works to visually unravel the collapse that seems immanent in the skyscraper's construction and instead suggests that what Petit did had the power to heal the imagined fragility of an urban landscape. As such, the film uses the sublimity of Petit's walk in order to stimulate the memory of 9/11 but hold its destructive effects at bay.

Man on Wire's representational modes transpose the modernist imagination of New York space onto a post-9/11 moment. At the outset of this book, I suggested that New York has a unique place in narrative both for its role in a burgeoning modernist culture and for the way that it articulates tensions generated by the clash of horizontal and vertical space. Crucially, *Man on Wire* works to recuperate Manhattan's potentially disjunctive spatiality as it is fixed by the 1920s "city symphonies" or "city films," which comprised a genre of avant-garde film conspicuously preoccupied with the idea of collapse and ruin in the metropolitan landscape. The genre is widely acknowledged as having been initiated by *Manhatta*, shot and produced in 1920 and 1921 by the painter Charles Scheeler and the photographer Paul Strand. *Manhatta* explores perspective from different points along a vertical axis, as well as making the skyscraper its central mode of depicting the city.[42] *Man on Wire* resurrects the vertical perspective of the city film, both distilling a sense of awe in the face of a vertical landscape and at the same time reframing the kind of disorienting camerawork of *Manhatta* in a way that attempts to stabilize the imminence of terrifying collapse. The final part of this chapter examines how *Man on Wire*'s visual strategies work to reconfigure the modernist city film in order to secure verticality as the source of the sublime, while also augmenting its central suggestion that vertical space forms an exception to the structures of everyday life.

Man on Wire's depiction of Petit's first entry into the North Tower describes his ascent through an elevator shaft. Petit portrays traveling up the skyscraper as an ascent to "heaven" while the camera looks down upon him; this narration is intercut with black-and-white shots

---

41. McCann, in interview with Johnston, "2009 National Book Award Winner."
42. See also Suarez, "City Space, Technology, Popular Culture"; Gartenberg and Westhelle, "NY, NY."

tracking an elevator's progress from a variety of angles, the camera at one point twisting from below to alongside the carriage. Petit describes "going all the way up" as "the darkness becomes gray" and the temporary freight elevator arrives "just a few floors below the roof." The verbal track denotes "a travel toward heaven" while the swaying camerawork and oscillating vertical perspective connote an especially precarious ascent. The black-and-white shots in the sequence focus on vertical lines and make use of a vertical perspective from below; the shots of Petit's interview come from above as opposed to the shots from underneath the elevator. The juxtaposition of vertical perspectives and the winding, disorienting movement of the elevator are indebted to the cinematography of *Manhatta*, but encode Petit's ascent as less ominous. Extending the suggestion that ascent is a precarious yet transcendent passage, *Man on Wire* then reframes a sequence from Robert Florey's 1929 city film, *Skyscraper Symphony*, with its own footage from 1974 panning up the length of the World Trade Center in a disorienting manner.[43]

In the sequence I refer to from *Skyscraper Symphony*, shots pan up from the base of buildings and twist around while directing the gaze upward, all the time growing faster and shakier to effect a pervasive sense of instability as languid shots give way to the increasing pace of editing and the heightened tempo of the soundtrack. The sequence in *Man on Wire* both visually alludes to *Skyscraper Symphony* and uses images of the now gone Twin Towers that bear a retroactive power secured by the audience's awareness of their subsequent collapse. In this sense, *Man on Wire*'s sequence must contend with the paradox of summoning nostalgia relating to the World Trade Center after 2001, a narrative difficulty also facing post-9/11 films set in New York before 2001 that were forced to include or excise the Twin Towers from their visual landscape, cognizant of the fact that both maneuvers would potentially remind the audience of the 2001 attacks.[44] The shots in *Man on Wire*, like the images of the buildings being constructed discussed at the outset of this chapter, suggest that the towers' collapse is immanent in their very construction. But they also, significantly, transcribe the visual syntax of the city film to prioritize Petit's ensuing artistic creation rather than a sense of looming destruction.

In *Man on Wire*, the skyscraper initially incites only terror; a shot directed toward the apex of the World Trade Center accompanies

---

43. *Manhatta* and *Skyscraper Symphony* are included in the DVD collection *Picturing a Metropolis: New York City Unveiled*.

44. Schneider, "Architectural Nostalgia"; Negra, "Structural Integrity," 51–52.

Heckel's narration of a "terrifying" walk through the city, "horrified" as he looks up at "the height of the Empire State Building." His legs start "to shake"; it is "terrifying"; then he realizes that the Twin Towers are "even higher." In a further attempt to work through the haunting tones of collapse, and to defuse the threat of extreme height, *Man on Wire*'s early inclusion of Heckel's perspective from the base of the Twin Towers is later reversed. This strategy transforms the act of looking up—which in *Skyscraper Symphony* is ominous and in 2001 was the terrifying experience of those watching the New York sky—from a narrative of fear into one of overwhelming beauty. Approximating Petit's view from the top of the North Tower, and referring to the vertiginous perspective authored most famously by Alfred Hitchcock in *Vertigo,* the film displaces the previous vantage point from the ground. In *Vertigo,* shots from above disorient the viewer in parallel with the characters within the narrative; the view in *Man on Wire* below works similarly, approximating Petit's vantage point in a visual reenactment for the audience.

Petit calls the tower on which he stands "a throne, the highest tower ever built by man," and the view of New York from this height an "unfathomable canyon" comprising his own "empire."[45] His words recall those of Michel de Certeau, who also grants vertical space exceptionality in his exposition of the strategies and tactics of everyday life. In *The Practice of Everyday Life* (1984), the idea of physically standing on top of the World Trade Center is both a primary example of, and the exception to, the spatial practices de Certeau describes:

> To be lifted to the summit of the World Trade Center is to be lifted out of the city's grasp. One's body is no longer clasped by the streets that turn and return it according to an anonymous law; nor is it possessed, whether as player or played, by the rumble of so many differences and the nervousness of New York traffic. When one goes up there, he leaves behind the mass that carries off and mixes up in itself any identity of authors or spectators. An Icarus flying above these waters, he can ignore the devices of Daedalus in mobile and endless labyrinths far below. His elevation transfigures him into a voyeur. It puts him as a distance. It transforms the bewitching world by which one was "possessed" into a text that lies before one's eyes. It allows one to read it, to be a solar Eye, looking down like a god.[46]

---

45. Petit, *To Reach the Clouds,* 178–79.
46. De Certeau, *Practice of Everyday Life,* 92.

Looking down from the summit of the World Trade Center delivers a transcendent experience precisely because it allows the viewer to "read" the world below, free from "an anonymous law." For Roland Barthes, too, the tower is exceptional: it is an object that "transgresses" the "habitual divorce of *seeing* and *being seen*"; "an object when we look at it" and "a lookout when we visit it." This same quality also makes the skyscraper function as symbolically unique: the tower's "sovereign circulation" between lookout and spectacle ensures that it "attracts meaning the way a lightning rod attracts thunderbolts."[47]

In *Man on Wire*, vertical ascent solicits an even more extraordinary act of "possession" than in de Certeau's or even Barthes's description: at the top of the towers, Petit experiences a transformative sensory ability, whether real or imagined, to hear and see those a quarter of a mile below. Feeling omniscient, Petit declares: "I can tell you, probably it's a lie, but to me it's not, I heard the crowd. I saw the crowd. I hear them murmur." *Man on Wire* also complicates what is a one-way experience in de Certeau's essay—looking down from a height affords power only to the person above—by implying that the transformative power of Petit's walk was reciprocal. In Moore's description, the sight is rendered "beyond anything you could imagine" and marked as a "magical" and "profound" event that caused Moore, as a spectator, to be overcome by "awe." These words explicitly unravel the "horror" associated with Heckel's perspective looking up and also imply that verticality affords not only the performer, but the witness, too, with a moment in which their entire comprehension of the world around them is altered. It transforms the idea of an arrested fall discussed earlier in this chapter, as well as at the outset of this book in relation to *Extremely Loud and Incredibly Close,* into the source of the sublime. It also conjures the re-mediation of the sublime moment as it was imagined by Kant as an experience stimulated by the confrontation with what is threatening without involving actual fear. Kant's definition is exactly the criterion that allows Petit's walk in *Man on Wire* to function as a symbol of transformation; because the film is a renarration of actual events, it can invoke the sublime without provoking the real fear that may have been felt by witnesses of the original performance in 1974.

The voiceover accompanying the shot looking down from the top of the World Trade Center in *Man on Wire* cements the discourses of exceptionality that surround the tower into a reiteration of the kind of deterministic fulfillment previously implied by Allix's suggestion that

---

47. Barthes, *Eiffel Tower,* 4–5 (emphases in original).

the World Trade Center was somehow "built especially" for Petit. In a breathless and revelatory tone, Petit now recalls: "And I had to make a decision of shifting my weight from one foot anchored to the building to the foot anchored on the wire." He then states that, while thinking that stepping onto the wire would "probably" be the "end of [his] life," that step was also something he could "not resist" and "didn't make any effort to resist." Petit's narration inscribes the walk as exceptional precisely because it is predetermined: he cannot resist, and he is not responsible, because the performance is guaranteed by a fatalistic structure. This deflection of an ethical paradigm recalls the fact that the capacity of the sublime to leave the beholder temporarily unable to understand what they are witnessing lies at the heart of a philosophical debate, one Christine Battersby describes as wondering whether the sublime can be "morally educative," and whether the "productive tension" of the sublime can successfully push "the mind" toward "an idea of a higher (hidden) order." Although reading 9/11 "in terms of the visual vocabulary of the sublime" is an almost automatic reaction to the idea of "a shattered landscape," the ethical and aesthetic dilemmas so prevalent after 9/11 that always develop when "political violence is viewed as 'art'" are as relevant as those when Burke and Kant were writing.[48]

Man on Wire both detaches the moment it narrates from an ethical context and invokes what would typically be called a sublime experience in a way that tries to work backward and somehow unravel the traumatic associations left by 9/11. The sublime is already paradoxical in its temporal figuration; it is an experience imagined as stepping out from the constraints of time and, in Kant's terms, also rests upon the retrospective assimilation of a profound experience into the parameters of everyday morality. Petit's autobiography explicitly mines this paradox by describing the walk as somehow outside time and also firmly embedded in a precise moment:

> All of a sudden, the density of the air is no longer the same. Jean-François ceases to exist. The facing tower is empty. The wheel of the elevator no longer turns.
> The horizon is suspended from east to west.
> New York no longer spreads its infinity. The murmur of the city dissolves into a squall whose chill and power I no longer feel.[49]

---

48. Battersby, *Sublime*, 24; 14; 21.
49. Petit, *To Reach the Clouds*, 163.

Petit's description conflates apparently competing ideas: the air is in flux, but Petit's companion abruptly stops existing; New York is fixed in a single moment but also described in terms of its previous "infinity." The present tense semantically suggests the immediacy of the moment while the words themselves allude to an eternal temporality. Thematically underlying the narration is the idea that the event described holds still, reverses, or even exceeds the limits of time. The sublime experience—in both *Man on Wire* and the text from which it comes—compounds the imagining of vertical space as an exception to cultural rules, civil parameters, and laws of physics. Where this idea coincides with a response—however oblique—to 9/11, the result is vertical imagery that both bears the weight of a catastrophic event and betrays a belief that this same imagery might alleviate some of the pain caused by that event.

The sublime moment, which has emerged from this book as a representative cohort of the vertical spectacle, is historically fraught by various and often conflicting ideas. While Kant distinguished terror in the sublime experience from real fear and imagined the feeling to be "morally educative," and his work perhaps provides the way for thinking through terror that "neither simply aestheticises that terror" nor "homogenises the response" to it, as Battersby points out, "the politics of the sublime in the face of real-life terror" are a problematic after-effect of conceptualizing the sublime.[50] Moreover, these politics remain unaddressed not only by the invocation of the sublime in post-9/11 texts such as *Man on Wire* but also by discourse that relegates the events of 11 September 2001 as an exceptional moment standing somehow outside time. This chapter has revealed the way in which *Man on Wire* is leveled at an unraveling of time, using retroactive representation to go backward and try to alleviate the traumas of the terror attacks. However, it is precisely this kind of retreat into the past that blocks a comprehensive analysis of the events of 2001. While there is not an overt political weight behind *Man on Wire* and the film uses nostalgic sentiment for a time before 9/11 in a decisively light-hearted manner, its retroactivity reveals a wider sense in which returning to an earlier moment potentially deflects attention away from a fuller understanding of the crossover between "the sublime and questions of political theory and value" voiced by Battersby.[51]

While maintaining a distinction between the inherent procedures of a text and the milieu in which it is received, I have argued that *Man on*

---

50. Battersby, *Sublime*, 24; 23; 43.
51. Ibid., 11.

Wire betrays an underlying desire to reimagine the events of 2001 by summoning those of 1974, using verticality to obliquely address the specifically vertical traumas and anxieties of 11 September 2001. By returning to a moment prior to the 2001 attacks, by initially alluding to criminal intent via the familiar grammar of the heist film only to expel these allusions from its plot, and by narratively repudiating the spectacular and hugely distressing fall from the top of the towers, *Man on Wire* attempts to secure its own parameters against the specter of 9/11: a phantom made near-viral in James Marsh's comments. Whereas the first part of this chapter reveals the way in which the film becomes entangled in the complexities of imagining Petit's walk free from the restraint of moral and cultural procedures, this latter section has suggested that the film's visual recapitulation of the modernist city film, while not immune from these same issues, allows verticality to operate within a frame of reference not confined to 9/11.

If the sublime is imagined as a moment that opens up a void in understanding that may never be assimilated, its invocation in post-9/11 narrative potentially facilitates a further turning away from the reality of the events and aftermath of 2001. Jacques Derrida asserted that on 11 September 2001 "'something' took place," but that "this very thing, the place and meaning of this 'event,' remains ineffable."[52] In Battersby's parsing, Derrida suggests that 9/11 involved a collision of temporalities and was not just a congregation of events: 9/11's impact comes not only from revealing "features in the past that we had previously 'forgotten' to see" but also from producing "a shock linked to anticipation and to other (possible) futures that are opened up by the sudden shift in perspective." As Battersby notes, Derrida's discussion of 9/11 suggests less a Heideggerian event, always beyond the limits of understanding, and more a Nietzschean understanding of the 2001 terror attacks, as beyond understanding precisely because of our historical time.[53] It nonetheless relies upon the vocabulary of the sublime to describe the rupturing force of the attacks.

The same is true in *Man on Wire*'s invocation of the sublime to talk about the events of 1974, and while the correspondence between the sublime in Petit's walk and the events of 11 September 2001 is problematic, it also uncovers the way in which the film's vertical imagery rests upon a sense of disrupted time. Framing the representation of verticality in

---

52. Derrida quoted in Borradori, *Philosophy in a Time of Terror*, 86.
53. Battersby, *Sublime*, 196; 194–96.

post-9/11 texts in terms of a relationship between the sublime moment and temporality does not provide an answer, or even an explicit response, to the kind of issues surrounding real-life terror Battersby articulates, but the expression of time somehow being reversed does uncover vertical imagery as a response to a specific historical moment. *Man on Wire*'s revivification of the city symphony allows the film to retroactively summon an ongoing discourse relating to the skyscraper's form, revealing a desire to draw together historical anxieties associated with verticality in order to diffuse them. The film does not merely recirculate the anxieties left by 9/11, but can be seen attempting to loosen the rigidity of destruction and trepidation as they are attached to the tall building. As such, the film's cultural retroactivity exposes a wider sense in which vertical imagery is directed at expressing temporal disruptions, here both in response to 9/11 and more generally as a way to gesture back to a longer view. By accentuating how vertical structures historically figure as the site of competing ideas, *Man on Wire* recalls Henry Adams's depiction of looking up at the dynamo that I read at the outset of this book; read together, these texts reveal the employment of vertical imagery to anticipate the power of human creative and scientific endeavor. Whether articulated through awe at the new dynamo or amazement at one man's death-defying act, both texts imply that verticality is underwritten by the expectation of a transformative moment in both the immediate spectator and a larger human community.

CODA

UP AND DOWN STORIES

Chris Ware's collection, *Building Stories* (2012), is not quite a graphic novel: it is a cluster of differently shaped graphic narratives presented in a large heavyweight box. Inside the box resides two hardbound comic books; several broadsheet paper comics; one paper booklet the height of one panel and width of several; and two more of a similar shape but with an accordion-style foldout structure. The box also contains a foldout hardboard akin to a board game; divided into four stacked portrait panels, the hardboard portrays the basement and three floors of a Chicago walk-up building in Humboldt Park, the primary setting of all the narratives in *Building Stories*. The various texts inside the box work to magnify and extend a specific formal property of comics typical of Ware's work—namely, that a single comics page can be appreciated by its reader in several different sequences. The individual texts that comprise *Building Stories* can be read in any order, and the lack of instruction as to which order is the "correct" one makes this an essential part of its narrative fluidity. The texts as a whole detail the lives of occupants whose residence on the three stories of the building coincide, and the collection was first published in weekly installments in the *New York Times Magazine*. Essays addressing the published newspaper version and subsequent hard collection have considered the narrative as part of Chris Ware's ongoing fascination with architecture, time, ordinariness, memory, melancholy, nostalgia, and canonical art.[1]

In *Building Stories,* the woman who lives on the third floor of the building that unites all the narratives wonders at the concept of house attics. She considers the "architectural precedent" of the attic, asking: "Why is it always the *attic* where we banish our past? Is it because, since it's always above us, it feels analogous to our minds?" Then she continues to ruminate, thinking about the act of looking up when we try to recall a memory, and wonders if there might somewhere be a culture that "imagined its memories residing somewhere other than the brain, like in the heart, or in their feet, and if they built their houses accordingly, storing things in the middle, or in the basement." This unnamed woman is the most central protagonist of the entire collection, an art school attendee who is now (by which I mean the "now" of the specific strip) struggling to find a career path. As a child, she had a leg amputated after an accident on a motorboat; she also has a weak heart, and her decision to live on the top floor was a conscious decision to ensure she would get daily exercise. Her thoughts about the symbolic emotional power accorded to the attic space in buildings comes in one of the two hardbound books included in *Building Stories*. It is a moment shot through with the kind of bleak humor endemic to all the narratives: later in the same book, the woman is divested of her meditations about attic space when she's informed by the man she will later marry—an architect—that buildings like the one she lives in "don't *have* attics." In fact, the basement is the storage space for the old woman who owns the building and lives on the first floor. The thoughts of the woman living on the third floor provide a moment of reflexivity in which the textual narrative of *Building Stories* circles back on its own means of mediation and representative strategies. They verbally describe a human behavior related to "looking up" to recall; to arranging a home around the relegation of those physical artifacts that represent a "past." They are words that simultaneously draw attention to the very process of construction involved in reading *Building Stories;* a process at a microlevel in which the panels on a comics page

---

1. Worden, "On Modernism's Ruins"; Godbey, "'Building Stories'"; Berman, "Imagining an Idiosyncratic Belonging"; Sattler, "Past Imperfect"; op de Beeck, "Found Objects"; Roeder, "Seeing Inside-out"; Freedman, "Chris Ware's Epiphanic Comics."

are built into a narrative, and, on the macrolevel, in which the individual building blocks of separate comics texts inside the box are combined to make one overarching narrative. The woman's thoughts work with the same metaphor essential to *Building Stories*: constructing a building is like constructing a narrative.

Building Stories is a text that operates at the juncture of various kinds of narrative form and concretizes the discussions of multigenre texts in this book. It might usefully be termed a multistoried graphic narrative in which *multistoried* is a deliberately doubled phrase that signals the relationship between narrative and architecture. As I discussed in the first chapter of this book, the doubled meaning of "story" as both narrative and the floor of a building originates in an etymological root that entwines storytelling with architecture. Ware's graphic work consistently poses the building and the comics page as deeply related; in both *Building Stories* and *Jimmy Corrigan* (2000), iterative images of tall buildings and other structures are frequently symbols that reflect upon, or condense, moments in a graphic narrative's storyline. Ware's own thoughts about comics *as* architecture are explicit and have been discussed in detail by scholars.[2] In this sense, *Building Stories* condenses what has emerged as a leitmotif of this book: the ways in which images of "up" and "down" preoccupy American narratives and, in particular, how these images become entwined with the tangibly vertical. *Building Stories* emphasizes the physical architecture of buildings in a manner comparable to *Extremely Loud and Incredibly Close*: that is, buildings are not only the prompt for proprioceptive experience, but provide an arch-metaphor that organizes a set of ideas and discourses. Ware's spatial metaphor specifically makes a vertical building analogous to a human body with memories located in the attic/brain, expressing a conflation of physical architecture and affective structures. Throughout, Ware's figures of constructed meaning and history come most strikingly through his visual representation of a vertical Chicago building "as" a human story.[3] The collection comprises several separate texts that can be unfolded like blueprints or made into vertical cardboard buildings. By making its prime metaphor the physical reality of the graphic art contained within the collection, *Building Stories* exhibits the methods that this book argues are essential to understanding verticality across the long twentieth century.

---

2. Ware, quoted in Raeburn, *Chris Ware*, 25. For discussions of Ware's equivalence between buildings and comics, see in particular Worden, "On Modernism's Ruins," and Godbey, "'Building Stories.'"

3. See also Worden, "On Modernism's Ruins"; Godbey, "'Building Stories.'"

The verticality of Ware's work also relies on the sense of waiting that determines the narratives studied in this book. In its representation of imminence, *Building Stories* makes verticality inextricable from questions of time and anxiety. Just as Winsor McCay's comics yoked vertical imagery with a ludic representation of time, so, too, does Ware's text annex verticality with a deliberately fluid chronology. At its most fundamental level, Ware's text disrupts the concept of a prescribed reading order. The various narratives that comprise *Building Stories* also marry explicitly vertical images with permutations of temporal moment; the text moves between the 1900s and the 1980s and, at one point, even imagines a scene 150 years in the future from the story's main action. The stories related by the text are marked by a profound sense of imminence; the characters who populate the narrative reflect repeatedly on the things that might or might not go wrong in their lives. In the narrative contained by one of the hardbound books, the female protagonist recalls the summer after her graduation, stating: "I had no idea what I was going to do with my life." The same text opens with four full pages of increasingly tense "silent" action in which the woman goes through the banalities of her everyday life. The images offer only onomatopoeic representations of the sounds of her showering, preparing food, and washing her clothes. Beginning in tranquility, the pages steadily accumulate tension in the expectation that something will go wrong. Nothing does.

The next page is a cutaway of the building, titled "Once Upon a Time," that returns to the history of the physical space the woman inhabits and disrupts the tight and silent narrative of the preceding pages in an unstructured double-spread in which the vertical building takes precedence. Although—or perhaps because—no disaster befalls the woman, the sense is that the vertical building houses an immanent power for unsettling the normal course of events. Waiting for the sky to fall is, in *Building Stories,* the sense that personal history might go awry at any time. My discussions in this book have returned, repeatedly, to the distortions and manipulations of temporality secured by various texts' formal strategies. Each chapter has produced the idea that there exists a sustained marriage between form and function in texts that emphasize time as a complement to vertical imagery. By revealing how the texts discussed problematize time's progression, the chapters of this book have shown verticality as an expression of waiting: often for something that may or may not happen. My reading of Ware's collection condenses some of the themes that have emerged throughout this book and asks how vertical metaphor creates relationships between narrative and time.

The woman's thoughts that comprise the quotation above have a reflexivity met by moments in several of the texts comprising *Building Stories* in which the building itself speaks, shown in curlicue words distinguished from the block lettering of the human characters' words. In one such example, the building is exposed in a vertical dissection, the rooms visible. Katherine Roeder suggests that this cutaway "demonstrates the ability of comics, like architecture, to represent past and present in a single moment, thereby dissolving the linear narrative."[4] For Nathalie op de Beeck, too, this cutaway panel is one that consummately attends to the concept of time; in her reading, the panel exemplifies Ware's effort to stimulate "our consciousness of ongoing time." The cutaway "pretends to bring all the history—all the 'building stories' that potentially *could* be told, yet hang suspended in the soon-to-be-forgotten past—simultaneously into spectatorial and spectral awareness."[5] Both these readings assume the conflation between comics-as-architecture and comics-as-time. What is missing is the recognition of vertical metaphor as the vehicle for both these concepts.

TEMPORAL DISTURBANCE AND THE SUBLIME

In the act of waiting for the sky to fall—an act that, in itself, is fraught by the question of what comes next—there is an implicit sense of anxiety. Waiting for the sky to fall is, after all, a phrase that denotes imminent disaster. In chapter 5, I suggested that the fall, as an event that might or might not end in fatality, condenses Caruth's idea of trauma as that which is in excess of predictability or comprehension.[6] It is this apprehension that repeatedly marks the vertical metaphor discussed in this book. Not being able to know is both the concept that secures vertical expression as apprehensive and relates it most deeply to the sublime experience. As I discussed in chapter 4, the American sublime as described by Rob Wilson refers to "an *identification* with vast space and power, however imaginary this sense," creating a literary genre that persists "as New World dialogue between 'Americas' past and future," providing a "purchase" on both diachronic and synchronic "dimensions of collective history."[7] The American sublime in this sense is a fascination with space, but space

---

4. Roeder, "Seeing Inside-out," 27.
5. Op de Beeck, "Found Objects," 825–26.
6. Caruth, *Unclaimed Experience*, 5–7; 73–90.
7. Wilson, *American Sublime*, 20; 28.

that is fundamentally *felt*; space that can be written about and imagined. The emergence of nuclear technology signified a challenge to this conceptualization in that it provided the means of complete annihilation; the sublime moment was simultaneously secured by the idea of the atomic bomb and negated by its possibility. It is this kind of paradox that recurs in texts that represent vertical images of sublimity.

    Furthermore, this book has revealed distortions or manipulations of temporality offered through texts' formal strategies and cleave form with function in narratives emphasizing time as a complement to vertical imagery. The interaction between textual and visual representations in McCay's dream-comics creates a dissonant temporality that reflects back upon the underlying tenor of the narrative within each strip. In Ginsberg's poetry, an accelerating pace and increasing enjambment reflect upon the speed of the space rocket depicted within "POEM Rocket." Vertical imagery and looping chronological structures or images that counter Western temporality inhabit Native American writing, joining together verticality with reformed ideas of what time means and how it is understood. While distinct, all these various textual characteristics are exemplary of the manner in which narratives, to various extents and in diverse ways, express a desire to rewrite the strictures of time through their formal manipulations. Where vertical imagery coalesces in moments of temporal disruption, verticality can be read as a specific response to what Mary Ann Doane calls "rationalized time," or the systematic ordering of time, and a means of considering how to represent time in light of this ordering. Rationalized time emerged at the beginning of the twentieth century from "a sea change in thinking about contingency, indexicality, temporality, and chance"; the cultural work of systemizing time is "in complicity with notions of the inevitability of a technologically induced historical process" and was later imagined by Walter Benjamin as "homogenous, empty time."[8] McCay's work indicates a dialogue between the underlying impact of systematized chronology and the changing shape of wider epistemic systems; furthermore, it exemplifies the ways in which vertical imagery frequently effects a sense of temporal disruptiveness that reflects back on the very nature of trying to order invisible concepts such as time.

    The way in which time becomes an essential part of vertical imagery extends from its conspicuous role in early twentieth-century texts into later texts such as Native American narratives that write against the grain of rationalized time. This response to temporality—in which systemized

---

8. Doane, *Emergence of Cinematic Time*, 4–7; Benjamin, *Illuminations*, 252.

time is actively interrogated—also underlies the emergence of the sublime as a privileged vocabulary through which verticality is represented. This book uncovers repetitive temporal paradoxes in the representation of the sublime experience when it is imagined as both a standing outside of, and also a moment embedded in, time, and because of this, the sublime emerges in line with Hegelian and Nietzschean readings of the Kantian sublime: here, "the potential of the sublime" is that it enables the subject "to transcend the limited framework of the space-time structures which shape our world."[9] Verticality is often annexed to the sublime as another way of unraveling the perceived rigidity of time itself. The relationship between the sublime spectacle and a destabilized sense of time embodies the way in which texts representing verticality are consistently preoccupied by an oscillation between concrete and abstract; between what can be sensorily perceived and what cannot.

Textual representations of vertical spectacle are often entrenched in the effect generated by the sensory experience of an overwhelming vertical event or structure, such as the skyscraper, the oil geyser, atomic bomb detonations, and the events of 9/11. However, the texts discussed in this book do not merely mediate what is visually vertical: they frequently either reimagine those same moments (exemplified by Ginsberg's poetic revision of dropped bombs and *Man on Wire*'s reframing of the new skyscraper of the 1920s city film and the lost skyscrapers after 9/11), or imagine entirely new ones (such as Silko's image of the Alternative Earth modules and McCay's exaggerated vertical world). This results in textual representations that may be directly prompted by a precise vertical event or structure but that inevitably become an element of the wider, and more complex, process of unraveling the route between concrete and abstract.

In this sense, vertical representations concerned with the visual effects of that verticality comprise a dialogue with Guy Debord's model of the spectacle as "a *weltanschauung*" (a worldview) that has been "actualized, translated into the material realm."[10] Allusions to the vertical spectacle of collapsing buildings might translate a moment into a tangible presence, or the "material realm." But vertical imagery reverses the coordinates of Debord's spectacle. Louis H. Sullivan's description of the skyscraper, in which he asserted that the skyscraper must hold "the force and power of altitude," and "must be every inch a proud and soaring thing, rising in

---

9. Battersby, *Sublime*, 10.
10. Debord, *Society of the Spectacle*, 13.

sheer exultation," makes what is physically vertical—the tall building—into a host for abstracted qualities such as "pride."[11] The competing demands of what both can and cannot be registered by the senses are pushed through the prism of verticality; the translation from conceptual to concrete involves a multidirectional circuit that also allows what is tangibly vertical to stand in for what is abstract.

Essential to this metaphorical project is the concept of the vertical frontier, which, throughout this book, has not simply described a historical trend but worked to unravel the implications of specific verticality from a variety of lineages. The vertical frontier outlined in the introduction of this book denotes the transference of emphasis from horizontal to vertical space after the closure of the frontier and its concomitant effect on the perceived lack of space lying west of the nation. The vertical frontier is the imagining of impressive skyscrapers in *Little Nemo* and their equally impressive ruin; McCay's vertical frontier expresses not only a sense of man's material progress but links this actual verticality with the more abstract revolutions in conceiving the human mind. In Sinclair's *Oil!*, the vertical frontier is about the exhaustion of horizontal space and the possibilities of technological and capitalist endeavor; the vertical frontier is male-oriented and supremely Euro-American. In Silko's *Almanac of the Dead*, by comparison, the vertical frontier is an imagined space through which indigenous American people might conceive liberation from horizontal exhaustion. While the prompt to imagining verticality is, as in Sinclair's work, a product of horizontal constraint, the Native American texts I have discussed in this book are born of a history of deracination and genocide rather than the Euro-American frontier spirit that is *Oil!*'s legacy. Ginsberg's poetry expands the concept of a vertical frontier set out by Native American texts by housing a transcendent and apparently limitless human experience in the sublime technology. In a similar way, the texts produced after 11 September 2001 that represent Petit's walk between the Twin Towers direct the power of the vertical frontier at a seemingly limitless experience. Finally, the vertical frontier in *Building Stories* is one that draws together the parameters of this book. Emerging from turn of the century temporality and extending into an imagined future, the narratives in Ware's text make the tall building that is its setting and main character into a symbol that speaks forward and backward.

The vertical frontier, then, signifies a metaphorical examination of temporality and its limits. And the act of waiting for the sky to fall that

---

11. Sullivan, "Tall Office Building Artistically Considered," 406.

comes with the idea of a vertical frontier holds within itself a concern for the visual significance of witnessing vertical spectacles. While Debord describes a postmodern "society of the spectacle," a preoccupation with visuality persists from the beginning of modernism into, and through, postmodernism. The visually encoded events of 11 September 2001, the televised wars in Afghanistan and Iraq, and the revelations of photographed torture at Abu Ghraib all furnished long-standing fears of visual fatigue with new urgency,[12] but the underlying question attached to the spectacle of disaster remains largely the same, and is a common and pervasive problem articulated by scholarship. This question asks: how can the viewer assimilate disaster while remaining sensitive to the atrocity represented by various texts, and without becoming somehow complicit in what is represented? By looking specifically at the mediation of vertical spectacle, this book reveals the unique role of vertical disaster in American narratives as a vocalization of, if not an answer to, this question. Because vertical disaster becomes attached to moments of temporal disruption summoned by the sublime experience, or invokes a particular textual conflict in tone or language, vertical spectacles come to reflect back on the very anxieties involved in the re-mediation of historical traumas such as 9/11 and atomic bomb detonations. In other texts, the conflicts attending images of vertical disaster indicate the force of new modes of scientific and technological knowledge. The idea of spectacle, read alongside questions of temporality, indicates verticality as the manifestation of an acknowledged relationship between, on one hand, what can be seen and, on the other, the consequences that sight has in terms of a supposed knowledge of the world.

VERTICALITY AT THE FISSURE OF WHAT IS KNOWN

Vertical imagery shelters a response to, variously, models that limit the remembering of traumatic national events like 9/11; conceptions of the internal workings of the mind; and understandings of time, the sublime, and social structure. Consistently, verticality returns to the idea that some kind of fissure in what was previously known has occurred and seeks out ways to represent that fissure, actively protest its consequences, or remedy its effects. Even when vertical imagery is ostensibly a reaction to a physically vertical event or structure such as *Little Nemo*'s and *Man on*

---

12. See Hirsch, "Collateral Damage," 1211.

Wire's representations of the skyscraper form, *Oil!*'s reaction to the sight of the oil geyser, or Ginsberg's depiction of the dropped atomic bomb, a response to the aftermath of that physically vertical moment is also tied into the text. Vertical imagery in these instances is not a fleeting representation of a particular spectacle but is embedded in each text's modes of examining the consequences of, respectively, new ways of conceiving consciousness, 9/11, new technologies of industrial extraction, and the scientific revolution precipitated by the ability to split the atom. In other texts, verticality is not stimulated by a particular physical event or structure but responds to specific models of thought. Sometimes these models already verticalize, as within racial or social hierarchies or the idea of a vertically stratified mind, and occasionally they are horizontal, such as cartographical systems or the idea of linear time. Although vertical imagery emerges as a reaction to apparently distinct moments of physical verticality or abstract taxonomies that may or may not be imagined as vertical, the uniting thread emerging from this book is that verticality is embedded in a reaction, and it is not a fleeting or insubstantial part of texts but essential to their overarching themes and formal maneuvers.

Vertical imagery holds a particular potency precisely because of the various dissonances its presence summons in a variety of contexts and discourses. In this way, verticality is frequently tautological; that is to say, vertical imagery both provides the means, and ensures the ends, of representing physical or epistemological disorientation. The route between what is materially experienced and what is abstractly conceived—the circuit between what is tangible and what is not—is a conversion mediated by disruptive forms and themes. It is precisely the disruption effected by these formal, thematic, and imagistic moments that secure what is articulated as disorienting or revolutionary. By making them coexistent, narratives often reveal both disorienting vertical imagery and temporal disruption as a response to the limits and constraints of forms of measurement. In a similar way, representations of vertical spectacle discussed in this book frequently uncover the paradoxes of the sublime experience. But, rather than indicate the redundancy of understanding vertical spectacle in terms of the sublime, these paradoxes ratify precisely what the sublime in its most basic sense is imagined to mean—a moment that cannot be held by previous understanding or the parameters of everyday life and, as such, at least temporarily exceeds the limits of a known world. Vertical imagery becomes embedded in the parameters of a sublime experience that, in spite of its historically diverse configurations, recurs as an expression of going beyond the routine and usual limits of everyday life.

Because of these tendencies, verticality can be read as a persistent reflection of a fissure in what was previously known.

Verticality is most usually mobilized in response to specific events; it regularly ushers in moments of temporal disruption; and it frequently aligns with the difficulties attached to representing disaster and enjoying the sublime moment. I argue that these should be read together rather than separately, as part of a wider process in which vertical imagery is resolutely directed at expressing an overturning of the known world, whether in terms of specific historical events or more general ways of understanding. The representation of visually realized vertical events reveals broader and often abstract cultural concepts, such as how we know what we know, how we measure progress and its desirability, how we use taxonomies to differentiate between groups and communities, and how we remember and renarrate catastrophic moments. In exploring the crossover between what is physically vertical, such as the skyscraper, and abstracted qualities that might be as polarized and simplistic as "good" and "bad," texts mobilize vertical imagery as a means of expressing the frisson generated by cultural ideas and other discourses they might try to displace. Verticality is not a uniform attempt to resolve this conflict. Rather, American narratives employ verticality to ask why particular modes of thought exist and why they are rigidly enforced, and to stimulate a dialogue between various discourses that might otherwise be kept separate.

BIBLIOGRAPHY

Abel, Marco. "Don DeLillo's 'In the Ruins of the Future': Literature, Images, and the Rhetoric of Seeing 9/11." *PMLA*, 118:5 (2003): 1236–50.

Achilles, Jochen. "The Subject-Object Paradigm: Conflict and Convergence in Theories of Landscape, Consciousness, and Techno-scape Since Emerson and Thoreau." In *Space in America: Theory, History, Culture*, edited by Klaus Benesch and Kerstin Schmidt, 53–90. Amsterdam: Rodopi, 2005.

Adams, Henry. *The Education of Henry Adams*. Oxford: Oxford University Press, 1999. First published 1907.

Adams, Rachel. *Continental Divides: Remapping the Cultures of North America*. Chicago and London: University of Chicago Press, 2009.

Adamson, Joni. "Indigenous Literatures, Multinaturalism, and Avatar: The Emergence of Indigenous Cosmopolitics." *American Literary History*, 24:1 (Spring 2012): 143–62.

Alexie, Sherman. *Flight*. London: Harvill Secker, 2008. First published 2007.

———. *Indian Killer*. London: Vintage, 1998. First published 1996.

———. *Reservation Blues*. New York: Atlantic Monthly Press, 1995.

Allen, Paula Gunn. *Off the Reservation: Reflections on Boundary-Busting, Border-Crossing Loose Canons*. Boston: Beacon Press, 1998.

———. "The Psychological Landscape of Ceremony." *American Indian Quarterly*, 5:1 (February 1979): 7–12.

———, ed. *Spider Woman's Granddaughters: Traditional Tales and Contemporary Writing by Native American Women*. London: Women's Press, 1990.

Anderson, Paul Thomas. *There Will Be Blood*, DVD. Miramax, 2007.

Anker, Elizabeth S. "Allegories of Falling and the 9/11 Novel." *American Literary History*, 23:3 (Fall 2011): 462–82.

Anzaldua, Gloria. *Borderlands/La Frontera: The New Mestiza*. San Francisco, CA: Aunt Lute Books, 1999. First published 1987.

Apter, Emily, and William Pietz, eds. *Fetishism as Cultural Discourse*. Ithaca and London: Cornell University Press, 1993.

Archuleta, Elizabeth. "Securing Our Nation's Roads and Borders or Re-Circling the Wagons? Leslie Marmon Silko's Destabilization of 'Borders.'" *Wicazo Sa Review*, 20:1 (Spring 2005): 113–37.

Arensberg, Mary, ed. *The American Sublime*. Albany, NY: State University of New York Press, 1986.

Arnold, Ellen L., ed. *Conversations with Leslie Marmon Silko*. Jackson: University Press of Mississippi, 2000.

Arthur, Arnold. *Radical Innocent: Upton Sinclair*. New York: Random House, 2006.

Atkinson, Ted. "'Blood Petroleum': *True Blood*, the BP Oil Spill, and Fictions of Energy /Culture." *Journal of American Studies*, 47:1 (February 2013): 213–29.

Auster, Paul. "On the High Wire." In *The Art of Hunger*, 249–60. New York: Penguin, 1997.

Austin, Thomas, and Wilma de Jong, eds. *Rethinking Documentary: New Perspectives, New Practices*. Maidenhead, UK: Open University Press, 2008.

Bachelard, Gaston. *Air and Dreams: An Essay on the Imagination of Movement*. Translated by Edith R. Farell and C. Frederick Farell. Dallas: Dallas Institute, 1988. First published 1943.

———. *The Poetics of Space*. Translated by Maria Jolas. Boston: Beacon Press, 1994. First published 1958.

Baer, Ulrich, ed. *110 Stories: New York Writes after September 11*. New York and London: New York University Press, 2002.

Bakhtin, Mikhail. *The Dialogic Imagination: Four Essays*. Edited by Michael Holquist. Austin: University of Texas Press, 1981. First published 1934.

Ball, David M., and Martha B. Kuhlman, eds. *The Comics of Chris Ware: Drawing Is a Way of Thinking*. Jackson: University Press of Mississippi, 2010. Ebook.

Ball, Eve, with Nora Henn and Lynda Sanchez. *Indeh, an Apache Odyssey*. Provo, UT: Brigham Young University Press, 1980.

Banita, Georgiana. *Plotting Justice: Narrative Ethics and Literary Culture after 9/11*. Lincoln and London: University of Nebraska Press, 2012.

Barrett, Ross, and Daniel Worden, "Oil Culture: Guest Editors' Introduction." *Journal of American Studies*, 46:2 (May 2012): 269–72.

Barthes, Roland. *The Eiffel Tower and Other Mythologies*. Translated by Richard Howard. New York: Farrar, Straus, and Giroux, 1979.

Battersby, Christine. *The Sublime, Terror and Human Difference*. London and New York: Routledge, 2007.

Baudelaire, Charles. "The Painter of Modern Life." In *Selected Writings on Art and Literature*, translated by P. E. Charvet, 390–435. London: Penguin, 1992.

Baudrillard, Jean. *The Spirit of Terrorism and Requiem for the Twin Towers*. London: Verso, 2002.

Bauerkemper, Joseph. "Narrating Nationhood: Indian Time and Ideologies of Progress." *Studies in American Indian Literatures*, 19:4 (Winter 2007): 27–53.

Beck, John. *Dirty Wars: Landscape, Power, and Waste in Western American Literature*. Lincoln: University of Nebraska Press, 2009.

Beigbeder, Frederic. *Windows on the World*. Paris: Grasset, 2003.

Bell, Robert C. "Circular Design in Ceremony." *American Indian Quarterly*, 5:1 (February, 1979): 47–62.

Bell, Virginia E. "Counter-Chronicling and Alternative Mapping in *Memoria Del Fuego* and *Almanac of the Dead*." *MELUS*, 25:3/4 (Autumn/Winter 2000): 5–30.

Benjamin, Walter. *Illuminations*. Edited and with an introduction by Hannah Arendt. London: Pimlico, 1999. First published 1950.

———. "On Some Motifs in Baudelaire." In *Illuminations*, edited and with an introduction by Hannah Arendt, 152–96. London: Pimlico, 1999. First published 1950.

Bercovitch, Sacvan. *The American Jeremiad*. Madison and London: University of Wisconsin Press, 1978.

Berger, James. "Falling Towers and Postmodern Wild Children: Oliver Sacks, Don DeLillo, and Turns against Language." *PMLA*, 120:2 (2005): 341–61.

Berman, Margaret Fink. "Imagining an Idiosyncratic Belonging: Representing Disability in Chris Ware's 'Building Stories.'" In *The Comics of Chris Ware: Drawing Is a Way of Thinking*, edited by David M. Ball and Martha B. Kuhlman, 191–205. Jackson: University Press of Mississippi, 2010.

Bevis, William. "Native American Novels: Homing In." In *Recovering the Word: Essays on Native American Literature*, edited by Brian Swann and Arnold Krupat, 580–620. Berkeley, Los Angeles, and London: University of California Press, 1987.

Biemann, Ursula, and Andrew Pendakis. "This Is Not a Pipeline: Thoughts on the Politico-Aesthetics of Oil." *Imaginations*, 3–2 (2012): 7–16.

Blackbeard, Bill. "The Greatest Strip That Ever Flopped." Introduction to Winsor McCay, *Little Nemo: Little Nemo in Slumberland and Little Nemo in the Land of Wonderful Dreams 1905–1914*, 5–7. China: Evergreen, 2000.

Bloodworth, William A. *Upton Sinclair*. Boston: G. K. Hall & Company, 1977.

Boehm, Scott. "Privatizing Public Memory: The Price of Patriotic Philanthropy and the Post-9/11 Politics of Display." *American Quarterly*, 58:4 (December 2006): 1147–70.

Boland, Kerry. "'We're All the Same People'?: The (A)Politics of the Body in Sherman Alexie's *Flight*." *Studies in American Indian Literatures*, 27:1 (Spring 2015): 70–95.

Borradori, Giovanna, ed. *Philosophy in a Time of Terror: Dialogues with Jürgen Habermas and Jacques Derrida*. Chicago and London: University of Chicago Press, 2003.

Boyer, Paul S. *By the Bomb's Early Light: American Thought and Culture at the Dawn of the Atomic Age*. Chapel Hill: University of North Carolina Press, 1994.

Bragard, Véronique, Christophe Dony, and Warren Rosenberg. Introduction to *Portraying 9/11: Essays on Representations in Comics, Literature, Film and Theatre*, edited by Véronique Bragard, Christophe Dony, and Warren Rosenberg, 1–9. Jefferson, NC, and London: McFarland & Company, 2011.

———, eds. *Portraying 9/11: Essays on Representations in Comics, Literature, Film and Theatre*. Jefferson, NC, and London: McFarland & Company, 2011.

Bredehoft, Thomas A. "Comics Architecture, Multidimensionality, and Time: Chris Ware's *Jimmy Corrigan: The Smartest Kid on Earth.*" *Modern Fiction Studies*, 52:4 (Winter 2006): 869–90.

Brenkman, John. "Freud the Modernist." In *The Mind of Modernism: Medicine, Psychology, and the Cultural Arts in Europe and America, 1880–1940*, edited by Mark S. Micale, 172–96. Stanford, CA: Stanford University Press, 2004.

Brigham, Ann. "Productions of Geographic Scale and Capitalist-Colonialist Enterprise in Leslie Marmon Silko's *Almanac of the Dead.*" *Modern Fiction Studies*, 50:2 (Summer 2004): 303–31.

Brill, A. A. "The Introduction and Development of Freud's Work in the United States." *American Journal of Sociology*, 45:3 (November 1939): 318–25.

Broderick, Mick. "Surviving Armageddon: Beyond the Imagination of Disaster." *Science Fiction Studies*, 20:3 (November 1993): 362–82.

Brooker, Peter. "Terrorism and Counternarratives: Don DeLillo and the New York Imaginary." *New Formations*, 57 (2005–6): 10–25.

Brooks, Peter. *Psychoanalysis and Storytelling.* Cambridge, MA, and Oxford: Blackwell, 1994.

———. *Reading for the Plot: Design and Intention in Narrative.* Oxford: Oxford University Press, 1984.

Brottman, Mikita. "The Fascination of the Abomination: The Censored Images of 9/11." In *Film and Television after 9/11*, edited by Wheeler Winston Dixon, 163–77. Carbondale: Southern Illinois University Press, 2004.

Brown, Bill. *The Material Unconscious: American Amusement, Stephen Crane, and the Economies of Play.* Cambridge, MA, and London: Harvard University Press, 1996.

Bruzzi, Stella. *New Documentary: A Critical Introduction.* London and New York: Routledge, 2000.

Buelens, Gert, Sam Durrant, and Robert Eaglestone. Introduction to *The Future of Trauma Theory: Contemporary Literary and Cultural Criticism*, edited by Gert Buelens, Sam Durrant, and Robert Eaglestone, 1–8. London and New York: Routledge, 2014.

Bukatman, Scott. *Matters of Gravity: Special Effects and Supermen in the 20th Century.* Durham, NC, and London: Duke University Press, 2003.

———. *The Poetics of Slumberland: Animated Spirits and the Animating Spirit.* Berkeley and London: University of California Press, 2012.

Burke, Edmund. *A Philosophical Enquiry into the Origin of Our Ideas of the Sublime and Beautiful.* Oxford: Basil Blackwell, 1987. First published 1756.

Burt, Ryan. "'Death Beneath This Semblance of Civilization': Reading Zitkala-Sa and the Imperial Imagination of the Romantic Revival." *Arizona Quarterly*, 66:2 (Summer 2010): 59–88.

Burt, Ryan E. "'Sioux Yells' in the Dawes Era: Lakota 'Indian Play,' The Wild West, and the Literatures of Luther Standing Bear." *American Quarterly*, 62:3 (September 2010): 617–37.

Burtynsky, Edward. *Oil.* New York: Steindl Photography International, 2009.

Butler, Judith. "Photography, War, Outrage." *PMLA*, 120:3 (2005): 822–27.

———. *Precarious Life: The Powers of Mourning and Violence.* London: Verso, 2006. First published 2004.

Campbell, James. *This Is the Beat Generation*. London: Secker & Warburg, 1999.

Campbell, W. Joseph. *Yellow Journalism: Puncturing the Myths, Defining the Legacies*. Westport, CT: Praeger Publishers, 2003.

Canemaker, John. *Winsor McCay: His Life and Art*. New York: Abbeville Press, 1987.

Carlson, Marla. "Looking, Listening, and Remembering: Ways to Walk New York after 9/11." *Theatre Journal*, 58 (2006): 395–416.

Carroll, Hamilton. "'Like Nothing in This Life': September 11 and the Limits of Representation in Don DeLillo's *Falling Man*." *Studies in American Fiction*, 40:1 (2013): 107–30.

———. "September 11 as Heist." *Journal of American Studies*, 45:4 (November 2011): 835–51.

Carroll, John. *Terror: A Meditation on the Meaning of September 11*. Melbourne: Scribe Publications, 2002.

Caruth, Cathy. *Literature in the Ashes of History*. Baltimore: Johns Hopkins University Press, 2013.

———. *Unclaimed Experience: Trauma, Narrative and History*. Baltimore: Johns Hopkins University Press, 1996.

Chelsea, David. "He Walks on Air—110 Stories High." In *9–11 Artists Respond, Volume I*, 10–11. Milwaukie, OR: Dark Horse Comics, 2002.

Chomsky, Noam. *9–11*. New York: Seven Stories, 2001.

Chute, Hillary. "Comics as Literature? Reading Graphic Narrative." *PMLA*, 123:2 (2008): 452–65.

———. "*Ragtime, Kavalier & Clay*, and the Framing of Comics." *Modern Fiction Studies*, 54:2 (Summer 2008): 268–301.

———. "Temporality and Seriality in Spiegelman's *In the Shadow of No Towers*." *American Periodicals*, 17:2 (2007): 228–44.

Chute, Hillary, and Marianne DeKoven. "Graphic Narrative." *Modern Fiction Studies*, 52:4 (Winter 2006): 767–82.

Codde, Philippe. "Philomela Revisited: Traumatic Iconicity in Jonathan Safran Foer's *Extremely Loud and Incredibly Close*." *Studies in American Fiction*, 35:2 (Fall 2007): 241–54.

Conte, Joseph M. "Don DeLillo's *Falling Man* and the Age of Terror." *Modern Fiction Studies*, 57:3 (Fall 2011): 559–83.

Cooper, Lydia R. "Beyond 9/11: Trauma and the Limits of Empathy in Sherman Alexie's *Flight*." *Studies in American Fiction*, 42:1 (Spring 2015): 123–44.

Cooper, Simon, and Paul Atkinson. "Graphic Implosion: Politics, Time, and Value in Post-9/11 Comics." In *Literature after 9/11*, edited by Ann Keniston and Jeanne Follansbee Quinn, 60–81. New York: Routledge, 2008.

Crafton, Donald. "McCay and Keaton: Colligating, Conjecturing, and Conjuring." *Film History: An International Journal*, 25:1–2 (2013): 31–44.

Craps, Stef. "Conjuring Trauma: The Naudet Brothers' 9/11 Documentary." *Canadian Review of American Studies*, 37:2 (2007): 183–204.

Cullingford, Elizabeth. "American Dreams: Emigration or Exile in Contemporary Irish Fiction?" *Éire-Ireland*, 49:3–4 (Fall/Winter 2014): 60–94.

Davis, Mike. *City of Quartz: Excavating the Future in Los Angeles.* London: Vintage, 1992.

———. "Dark Raptures: A Consumers' Guide to the Destruction of Los Angeles." *Grand Street,* 59 (Winter 1997): 6–17.

———. *Ecology of Fear: Los Angeles and the Imagination of Disaster.* New York: Vintage, 1998.

———. "Isle of California." *Grand Street,* 71 (Spring 2003): 103.

Dawes, Henry. "Have We Failed with the Indian?" *Atlantic Monthly,* 84:502 (August 1899): 280–85.

De Certeau, Michel. *The Practice of Everyday Life.* Translated by Steven Rendall. Berkeley, Los Angeles, and London: University of California Press, 1988.

Debord, Guy. *The Society of the Spectacle.* New York: Zone, 1994. First published 1967.

DeGroot, Gerard J. *The Bomb: A History of Hell on Earth.* London: Pimlico, 2005.

Deleuze, Gilles. *Cinema 1: The Movement Image.* Translated by Hugh Tomlinson and Barbara Habberjam. London: Athlone, 1986.

DeLillo, Don. *Falling Man.* London: Picador, 2007.

———. "In the Ruins of the Future: Reflections on Terror and Loss in the Shadow of September." *Harper's Magazine,* 303 (2001): 33–40.

Deloria Jr., Vine. *God Is Red: A Native View of Religion.* Golden, CO: North American Press, 1994. First published 1973.

Deloria, Philip J. *Playing Indian.* New Haven, CT, and London: Yale University Press, 1998.

Dennett, Andrea Stulman, and Nina Warnke. "Disaster Spectacles at the Turn of the Century." *Film History,* 4:2 (1990): 101–11.

Dennis, Helen May. *Native American Literature: Towards a Spatialized Reading.* London and New York: Routledge, 2007.

Denzin, Norman K., and Yvonna S. Lincoln, eds. *9/11 in American Culture.* New York and Oxford: Altamira Press, 2003.

DeRosa, Aaron. "September 11 and Cold War Nostalgia." In *Portraying 9/11: Essays on Representations in Comics, Literature, Film and Theatre,* edited by Véronique Bragard, Christophe Dony, and Warren Rosenberg, 58–72. Jefferson, NC, and London: McFarland & Company, 2011.

Dixon, Wheeler Winston, ed. *Film and Television after 9/11.* Carbondale: Southern Illinois University Press, 2004.

Doane, Mary Ann. *The Emergence of Cinematic Time: Modernity, Contingency, the Archive.* Cambridge, MA, and London: Harvard University Press, 2002.

Donald, James, Anne Friedberg, and Laura Marcus, eds. *Close Up 1927–1933: Cinema and Modernism.* London: Cassell, 1998.

Duerfahrd, Lance. "A Scale That Exceeds Us: The BP Gulf Spill Footage and Photographs of Edward Burtynsky." *Imaginations,* 3-2 (2012): 115–29.

Durham, Philip, and Everett L. Jones, eds. *The Frontier in American Literature.* New York: Odyssey Press, 1969.

Duvall, John N. "Witnessing Trauma: *Falling Man* and Performance Art." In *Don DeLillo: Mao II, Underworld, Falling Man,* edited by Stacey Olster, 152–68. London and New York: Continuum, 2011.

Duvall, John N., and Robert P. Marzec. "Narrating 9/11." *Modern Fiction Studies*, 57:3 (Fall 2011): 381–400.

Edkins, Jenny. "Time, Personhood, Politics." In *The Future of Trauma Theory: Contemporary Literary and Cultural Criticism*, edited by Gert Buelens, Sam Durrant, and Robert Eaglestone, 127–39. London and New York: Routledge, 2014.

———. *Trauma and the Memory of Politics*. Cambridge: Cambridge University Press, 2003.

Eickhoff, Friedrich-Wilhelm. "On *Nachträglichkeit:* The Modernity of an Old Concept." *International Journal of Psychoanalysis*, 87 (2006): 1453–69.

Eisner, Will. *Comics and Sequential Art: Principles and Practices from the Legendary Cartoonist*. New York and London: W. W. Norton & Company, 2008. First published 1985.

Ekaku, Hakuin. "The Text of the Heart Sutra." In *Zen Words for the Heart: Hakuin's Commentary on the Heart Sutra*, translated by Norman Waddell. Boston and London: Shambhala Publications, 1996. Kindle edition.

Elkind, Sarah S. "Oil in the City: The Fall and Rise of Oil Drilling in Los Angeles." *Journal of American History*, 99:1 (June 2012): 82–90.

Elliot, Michael A. "Indians, Incorporated." *American Literary History*, 19:1 (Spring 2007): 141–58.

Epstein, Rob, and Jeffrey Freidman. *Howl*, DVD. Werc Werk Works: 2010.

Erdoes, Richard, and Alfonso Ortiz. *American Indian Trickster Tales*. New York: Viking, 1998.

Erdrich, Louise. *The Bingo Palace*. London: Harper Perennial, 2004.

Eyre, Chris. *Smoke Signals*, DVD. Miramax, 1998.

Falconer, Rachel. *Hell in Contemporary Literature: Western Descent Narratives since 1945*. Edinburgh: Edinburgh University Press, 2007.

Fanuzzi, Robert, and Michael Wolfe, eds. *Recovering 9/11 in New York*. Newcastle upon Tyne, UK: Cambridge Scholars, 2014.

Felman, Shoshana, ed. *Literature and Psychoanalysis: The Question of Reading: Otherwise*. Baltimore and London: Johns Hopkins University Press, 1982.

———. "To Open the Question." In *Literature and Psychoanalysis: The Question of Reading: Otherwise*, 5–10. Baltimore and London: Johns Hopkins University Press, 1982.

Ferguson, Harvie. *The Lure of Dreams: Sigmund Freud and the Construction of Modernity*. London and New York: Routledge, 1996.

Fiedler, Leslie A. *The Return of the Vanishing American*. London: Jonathan Cape, 1968.

Fischl, Eric. *Eric Fischl Online Photographic Collection*. http://www.ericfischl.com/index.html.

Foer, Jonathan Safran. *Extremely Loud and Incredibly Close*. London: Penguin, 2005.

Freedman, Ariela. "Chris Ware's Epiphanic Comics." *Partial Answers: Journal of Literature and the History of Ideas*, 13:2 (June 2015): 337–58.

Freud, Sigmund. "Anxiety." In *Introductory Lectures on Psycho-Analysis*, 328–43. London: George Allen and Unwin, 1922.

———. "An Autobiographical Study." In *The Standard Edition of the Complete Works of Sigmund Freud*, edited by James Strachey, Vol. 20, 7–70. London: Vintage, 2001. First published 1925.

———. *Beyond the Pleasure Principle*. Edited by James Strachey. London and New York: W. W. Norton, 1961. First published 1919.

———. "Creative Writers and Daydreaming." In *The Standard Edition of the Complete Psychological Works of Sigmund Freud*, edited by James Strachey, Vol. 9, 142–53. London: Hogarth, 1959. First published 1907.

———. "Difficulties and Preliminary Approach to the Subject." In *Introductory Lectures on Psycho-Analysis*, 67–81. London: George Allen and Unwin, 1922.

———. "The Dissection of the Psychical Personality." In *The Standard Edition of the Complete Psychological Works of Sigmund Freud*, edited by James Strachey, Vol. 22, 57–80. London: Vintage, 2001. First published 1933.

———. "The Dream-Work." In *Introductory Lectures on Psycho-Analysis*, 143–54. London: George Allen and Unwin, 1922.

———. "Fetishism." In *The Standard Edition of the Complete Works of Sigmund Freud*, edited by James Strachey, Vol. 21, 149–58. London: Vintage, 2001.

———. *Group Psychology and the Analysis of the Ego*. Translated by James Strachey. London: Hogarth Press, 1948. First published 1922.

———. "Inhibitions, Symptoms and Anxiety." In *The Standard Edition of the Complete Works of Sigmund Freud*, edited by James Strachey, Vol. 20, 87–174. London: Vintage, 2001. First published 1925.

———. *The Interpretation of Dreams*. Edited by James Strachey. London: Penguin, 1991. First published 1900.

———. *Introductory Lectures on Psycho-Analysis*. London: George Allen and Unwin, 1922.

———. *Jokes and Their Relation to the Unconscious*. Edited by James Strachey. Harmondsworth, UK: Penguin, 1976. First published 1905.

———. "Letter 75, November 14th, 1897." In *The Origins of Psycho-Analysis: Letters to Wilhelm Fliess, Drafts and Notes, 1887–1902*, edited by Marie Bonaparte, Anna Freud, and Ernst Kris, 229–35. London: Imago Publishing Company, 1954.

———. "Manifest Content and Latent Thoughts." In *Introductory Lectures on Psycho-Analysis*, 94–104. London: George Allen and Unwin, 1922.

———. "The Neuro-Psychoses of Defence." In *The Standard Edition of the Complete Psychological Works of Sigmund Freud*, edited by James Strachey, 41–67. London: Hogarth, 1959. First published 1894.

———. *New Introductory Lectures on Psycho-Analysis*. Edited by James Strachey. New York and London: W. W. Norton & Company, 1989. First published 1933.

———. *The Origins of Psycho-Analysis: Letters to Wilhelm Fliess, Drafts and Notes, 1887–1902*. Edited by Marie Bonaparte, Anna Freud, and Ernst Kris. London: Imago Publishing Company, 1954.

———. "Project for a Scientific Psychology." In *The Origins of Psycho-Analysis: Letters to Wilhelm Fliess, Drafts and Notes, 1887–1902*, edited by Marie Bonaparte, Anna Freud, and Ernst Kris, 347–445. London: Imago Publishing Company, 1954. First published 1897.

———. "The Question of Lay Analysis." In *The Standard Edition of the Complete Works of Sigmund Freud*, edited by James Strachey, Vol. 20, 183–258. London: Vintage, 2001. First published 1926.

———. *The Standard Edition of the Complete Psychological Works of Sigmund Freud.* Edited by James Strachey, Vol. 1, 1886–1899. London: Hogarth Press and Institute of Psychoanalysis, 1966.

———. "Symbolism in Dreams." In *Introductory Lectures on Psycho-Analysis,* 125–42. London: George Allen and Unwin, 1922.

———. "Typical Dreams." In *The Interpretation of Dreams,* 185–210. Oxford: Oxford University Press, 1999.

———. *The Unconscious.* London: Penguin, 2005. First published 1915.

Friedberg, Anne. "Reading Close Up, 1927–1933." In *Close Up 1927–1933: Cinema and Modernism,* edited by James Donald, Anne Friedberg, and Laura Marcus, 1–26. London: Cassell, 1998.

Frongia, Antonello. "The Shadow of the Skyscraper: Urban Photography and Metropolitan Irrationalism in the Stieglitz Circle." In *The American Skyscraper: Cultural Histories,* edited by Roberta Moudry, 217–33. Cambridge: Cambridge University Press, 2005.

Frost, Laura. "Still Life: 9/11's Falling Bodies." In *Literature after 9/11,* edited by Ann Keniston and Jeanne Follansbee Quinn, 180–206. New York and London: Routledge, 2008.

Furlan, Laura M. "Remapping Indian Country in Louise Erdrich's *The Antelope Wife.*" *Studies in American Indian Literatures,* 19:4 (Winter 2007): 54–76.

Gardner, Jared. *Projections: Comics and the History of Twenty-First-Century Storytelling.* Stanford, CA: Stanford University Press, 2012.

Gartenberg, Jon, and Alex Westhelle. "NY, NY: A Century of City Symphony Films." *Framework: The Journal of Cinema and Media,* 55:2 (Fall 2014): 248–76.

Gay, Peter. *Freud: A Life for Our Time.* London: Macmillan, 1988.

Gerstein, Mordicai. *The Man Who Walked between the Towers.* New York: Square Fish, 2007. First published 2003.

Gessen, Keith. "Horror Tour." *New York Review of Books,* 52:14 (22 September 2005): 68–72.

Gherovici, Patricia. "Freud's Dream of America." In *The Dreams of Interpretation: A Century down the Royal Road,* edited by Catherine Liu, John Mowitt, Thomas Pepper, and Jakki Spicer, 39–53. Minneapolis and London: University of Minnesota Press, 2007.

Ginsberg, Allen. *Allen Verbatim: Lectures on Poetry, Politics, Consciousness.* Edited by Gordon Ball. New York: McGraw-Hill Book Company, 1974.

———. *Collected Poems: 1947–1997.* New York: HarperCollins, 2006.

———. *Deliberate Prose: Selected Essays 1952–1995.* London: Penguin, 2000.

———. *Empty Mirror: Early Poems.* New York: Corinth Books, 1961.

———. *Howl: A Graphic Novel.* Animation art by Eric Drooker. London: Penguin, 2010.

———. *Indian Journals March 1962–May 1963.* San Francisco, CA: Dave Haselwood Books and City Light Books, 1974. First published 1970.

———. *Journals: Early Fifties, Early Sixties.* Edited by Gordon Ball. New York: Random House, 1978.

———. *Journals: Mid-Fifties:1954–1958.* Edited by Gordon Ball. London: Penguin, 1996.

———. "Notes to *Plutonian Ode*." In *Collected Poems: 1947–1997*, 803–7. New York: HarperCollins, 2006.

———. "Poetry, Violence, and the Trembling Limbs, or, Independence Day Manifesto." In *Deliberate Prose: Selected Essays, 1952–1995*, 3–5. London: Penguin, 2000. First published 1959.

———. *Spontaneous Mind: Selected Interviews, 1958–96*. Edited by David Carter. New York: Perennial, 2002.

Godbey, Matt. "'Building Stories,' Gentrification, and the Lives of/in Houses." In *The Comics of Chris Ware: Drawing Is a Way of Thinking*, edited by David M. Ball and Martha B. Kuhlman, 121–34. Jackson: University Press of Mississippi, 2010.

Goodison, Bruce. *Flight 93: The Flight That Fought Back*, DVD. Brook Lapping Productions, 2005.

Gordon, Ian. *Comic Strips and Consumer Culture: 1890–1945*. Washington and London: Smithsonian Institution Press, 1998.

Graham, Lloyd. "Blondin, the Hero of Niagara." *American Heritage Magazine*, August 1958.

Gray, Richard. *After the Fall: American Literature Since 9/11*. Oxford: Wiley-Blackwell, 2011.

———. "Open Doors, Closed Minds: American Prose Writing at a Time of Crisis." *American Literary History*, 21:1 (Spring 2009): 128–48.

Greenberg, Judith, ed. *Trauma at Home: After 9/11*. Lincoln and London: University of Nebraska Press, 2003.

Greengrass, Paul. *United 93*, DVD. Universal Pictures, 2006.

Groensteen, Thierry. *The System of Comics*. Jackson: University Press of Mississippi, 2007. First published 1999.

Groves, General Leslie. "Memorandum for the Secretary of War." 18 July 1945. http://www.atomicarchive.com/Docs/Trinity/Groves.shtml.

Gunn, Joshua. "Mourning Speech: Haunting and the Spectral Voices of Nine-Eleven." *Text and Performance Quarterly*, 24:2 (April 2004): 91–114.

Gunning, Tom. "Never Seen This Picture Before: Muybridge and Multiplicity." In *Time Stands Still: Muybridge and the Instantaneous Photograph Movement*, edited by Phillip Prodger, 222–72. New York: Oxford University Press, 2003.

Habermas, Jürgen. "Modernity—An Incomplete Project." In *The Norton Anthology of Theory and Criticism*, edited by Vincent B. Leitch, 1748–59. New York and London: W. W. Norton & Company, 2001. First published 1980.

Hadda, Janet. "Ginsberg in Hospital." *American Imago*, 65:2 (Summer 2008): 229–59.

Hale, Jr., Nathan G. *Freud and the Americans: The Beginnings of Psychoanalysis in the United States, 1876–1917*. New York: Oxford University Press, 1971.

Harding, Jeremy. "The Deaths Map: The War on America's Southern Border." *London Review of Books*, 20 October 2011, 7–13.

Harris, Leon. *Upton Sinclair: American Rebel*. New York: Thomas Y. Cromwell, 1975.

Hartman, Anne. "Confessional Counterpublics in Frank O'Hara and Allen Ginsberg." *Journal of Modern Literature*, 28:4 (Summer 2005): 40–56.

Haslam, Gerald. "Literary California: 'The Ultimate Frontier of the Western World.'" *California History*, 68:4 (Winter 1989): 188–95.

Hassler-Forest, Dan A. "From Flying Man to Falling Man: 9/11 Discourse in Superman Returns and Batman Begins." In *Portraying 9/11: Essays on Representations in Comics, Literature, Film and Theatre,* edited by Véronique Bragard, Christophe Dony, and Warren Rosenberg, 134–46. Jefferson, NC, and London: McFarland & Company, 2011.

———. "From Trauma Victim to Terrorist: Redefining Superheroes in Post-9/11 Hollywood." In *Comics as a Nexus of Cultures: Essays on the Interplay of Media, Disciplines and International Perspectives,* edited by Mark Berninger, Jochen Ecke, and Gideon Haberkorn, 33–44. Jefferson, NC: McFarland & Company, 2010.

Hatfield, Charles. *Alternative Comics: An Emerging Literature.* Jackson: University Press of Mississippi, 2005.

Hauwerwas, Stanley, and Frank Lentricchia, eds. *Dissent from the Homeland: Essays after September 11.* Durham, NC, and London: Duke University Press, 2003.

Hazard, Lucy. *The Frontier in American Literature.* New York: F. Ungar, 1961. First published 1927.

Heller, Dana, ed. *The Selling of 9/11: How a National Tragedy Became a Commodity.* New York and Hampshire, UK: Palgrave MacMillan, 2005.

Herms, Dieter, ed. *Upton Sinclair: Literature and Social Reform.* Frankfurt am Main, Bern, New York, and Paris: Peter Lang, 1990.

Hine, Lewis Wickes. *America and Lewis Hine: Photographs 1904–1940.* New York: Aperture, 1977.

Hirsch, Marianne. "Collateral Damage." *PMLA,* 119:5 (October 2004): 1209–15.

Holm, Sharon. "The 'Lie' of the Land: Native Sovereignty, Indian Literary Nationalism, and Early Indigenism in Leslie Marmon Silko's *Ceremony.*" *American Indian Quarterly,* 32:3 (Summer 2008): 243–74.

Honeyman, Susan. "Gastronomic Utopias: The Legacy of Political Hunger in African American Lore." *Children's Literature,* 38 (2010): 44–63.

Hornung, Alfred. "Literary Conventions and the Political Unconscious in Upton Sinclair's Work." In *Upton Sinclair: Literature and Social Reform,* edited by Dieter Herms, 24–38. Frankfurt am Main, New York, and Paris: Peter Lang, 1990.

Hoxie, Frederick E. *A Final Promise: The Campaign to Assimilate the Indians, 1880–1920.* Cambridge, New York, and Melbourne: University of Cambridge Press, 1984.

Hoxie, Frederick E., and Peter Iverson, eds. *Indians in American History.* Wheeling, IL: Harlan Davidson, 1998.

Huehls, Mitchum. "Foer, Spiegelman, and 9/11's Timely Traumas." In *Literature after 9/11,* edited by Ann Keniston and Jeanne Follansbee Quinn, 42–59. New York and London: Routledge, 2008.

Hughes, Robert. *The Shock of the New: Art and the Century of Change.* London: Thames and Hudson, 1991.

Huhndorf, Shari Michelle. *Mapping the Americas: The Transnational Politics of Contemporary Native Culture.* Ithaca, NY, and London: Cornell University Press, 2009.

———. "Picture Revolution: Transnationalism, American Studies, and the Politics of Contemporary Native Culture." *American Quarterly,* 61:2 (June 2009): 359–81.

Hungerford, Amy. "Postmodern Supernaturalism: Ginsberg and the Search for a Supernatural Language." *Yale Journal of Criticism,* 18:2 (Fall 2005): 269–98.

Irom, Bimbisar. "Alterities in a Time of Terror: Notes on the Subgenre of the American 9/11 Novel." *Contemporary Literature*, 53:3 (Fall 2012): 517–47.

Jackson, Brian. "Modernist Looking: Surreal Impressions in the Poetry of Allen Ginsberg." *Texas Studies in Literature and Language*, 52:3 (Fall 2010): 298–323.

Jacobson, Sid, and Ernie Colon. *The 9/11 Report: A Graphic Adaptation*. New York: Hill and Wang, 2006.

James, Henry. *The American Scene*. New York: Charles Scribner's Sons, 1946. First published 1907.

James, William. "The Hidden Self." *Scribner's Magazine*, 7:3 (March 1890): 361–73.

———. "The Stream of Thought." In *The Principles of Psychology*, Vol. 1, 224–90. New York: Cosimo, 2007. First published 1890.

JanMohamed, Abdul R. "The Economy of Manichean Allegory: The Function of Racial Difference in Colonialist Literature." In *"Race," Writing, and Difference*, edited by Henry Louis Gates, Jr., 78–106. Chicago: Chicago University Press, 1986.

Jarman, Michelle. "Exploring the World of the Different in Leslie Marmon Silko's *Almanac of the Dead*." *MELUS*, 31:3 (Fall 2006): 147–68.

Jefferson, Thomas. *Notes on the State of Virginia*. Chapel Hill: University of North Carolina Press, 1955. First published 1794.

Johnston, Bret Anthony. "2009 National Book Award Winner Fiction: Interview with Colum McCann." 2009. http://www.nationalbook.org/nba2009_f_mccann_interv.html.

Julius, Anthony. *Transgressions: The Offences of Art*. London: Thames and Hudson, 2002.

Junod, Tom. "The Falling Man." *Esquire*, September 2003. http://classics.esquire.com/the-falling-man/.

———. "The Falling Man, Ten Years On: Surviving the Fall." *Esquire*, 9 September 2011. http://www.esquire.com/news-politics/news/a10891/the-falling-man-10-years-later-6406030/.

Kamin, Blair. *Terror and Wonder: Architecture in a Tumultuous Age*. Chicago and London: Chicago University Press, 2010.

Kant, Immanuel. *The Critique of Judgement*. Translated by J. C. Meredith. Oxford: Clarendon Press, 1952. First published 1790.

Kaplan, Amy. *The Anarchy of Empire in the Making of U.S. Culture*. Cambridge, MA: Harvard University Press, 2002.

———. "Homeland Insecurities: Reflections on Language and Space." *Radical History Review*, 85 (Winter 2003): 82–93.

———. "Where Is Guantanamo?" *American Quarterly*, 57:3 (September 2005): 831–58.

Kaplan, Amy, and Donald E. Pease, eds. *Cultures of United States Imperialism*. Durham, NC: Duke University Press, 1993.

Kaplan, E. Ann. *Trauma Culture: The Politics of Terror and Loss in Media and Literature*. New Brunswick, NJ, and London: Rutgers University Press, 2005.

———. "'Wounded New York': Rebuilding and Memorials to 9/11." In *Trauma Culture: The Politics of Terror and Loss in Media and Literature*, 136–47. New Brunswick, NJ, and London: Rutgers University Press, 2005.

Kauffman, Linda S. "Bodies in Rest and Motion in *Falling Man*." In *Don DeLillo: Mao II, Underworld, Falling Man,* edited by Stacey Olster, 135–51. London and New York: Continuum, 2011.

———. "The Wake of Terror: Don DeLillo's 'In the Ruins of the Future,' 'Baadermeinhof,' and *Falling Man*." *Modern Fiction Studies,* 54:2 (Summer 2008): 353–77.

Kaufman, Eleanor. "Falling from the Sky: Trauma in Perec's *W* and Caruth's *Unclaimed Experience*." *Diacritics,* 28:4 (Winter 1998): 44–53.

Keiser, Albert. *The Indian in American Literature.* New York: Oxford University Press, 1933.

Keniston, Ann. "'Not Needed, Except as Meaning': Belatedness in Post-9/11 American Poetry." *Contemporary Literature,* 52:4 (Winter 2011): 658–83.

Keniston, Ann, and Jeanne Follansbee Quinn, eds. *Literature after 9/11.* New York and London: Routledge, 2008.

Kennedy, Martha A. "Drawing (Cartoons) from Artistic Traditions." *American Art,* 22:1 (Spring 2008): 10–15.

Kent, Alicia A. "'You Can't Run Away Nowadays': Redefining Modernity in D'Arcy McNickle's *The Surrounded*." *Studies in American Indian Literatures,* 20:2 (Summer 2008): 22–46.

Kerkhoff, Ingrid. "Wives, Blue Blood Ladies, and Rebel Girls: A Closer Look at Upton Sinclair's Females." In *Upton Sinclair: Literature and Social Reform,* edited by Dieter Herms, 176–94. Frankfurt am Main, Bern, New York, and Paris: Peter Lang, 1990.

Kern, Stephen. *The Culture of Time and Space, 1880–1918.* Cambridge, MA: Harvard University Press, 2001. First published 1983.

Kessler, Fritz C., and Frank Jacobs. *Mapping America: Exploring the Continent.* London: Black Dog, 2010.

Kimmage, Michael. "Atomic Histiography." *Reviews in American History,* 38:1 (March 2010): 145–52.

Klein, Naomi. "The Rise of the Fortress Continent." *The Nation,* 3 February 2003. http://www.thenation.com/article/rise-fortress-continent.

Koolhaas, Rem. *Delirious New York: A Retroactive Manifesto for Manhattan.* New York: Monacelli Press, 1994.

Kramer, Jane. *Allen Ginsberg in America.* New York: Fromm International Publishing, 1997. First published 1968.

Kripal, Jeffrey John. "Reality against Society: William Blake, Antinomianism, and the American Counterculture." *Common Knowledge,* 13:1 (Winter 2007): 98–112.

Kroes, Rob. "The Ascent of the Falling Man: Establishing a Picture's Iconicity." *Journal of American Studies,* 45:4 (November 2011), special electronic content: 1–20.

Krupat, Arnold. *Ethnocriticism: Ethnography, History, Literature.* Berkeley: University of California Press, 1992.

———. *New Voices in Native American Literary Criticism.* Washington, DC: Smithsonian Institution Press, 1993.

———. *Red Matters: Native American Studies.* Philadelphia: University of Pennsylvania Press, 2002.

———. *The Turn to the Native: Studies in Criticism and Culture.* Lincoln: University of Nebraska Press, 1996.

———. *The Voice in the Margin: Native American Literature and the Canon.* Berkeley, Los Angeles, and Oxford: University of California Press, 1989.

Lacan, Jacques. *Ecrits: A Selection.* London: Routledge, 2001. First published 1966.

Landau, Sarah Bradford, and Carl W. Condit. *The Rise of the New York Skyscraper, 1865–1913.* New Haven, CT, and London: Yale University Press, 1999.

Laplanche, Jean. "Notes on Afterwardsness." In *Essays on Otherness / Jean Laplanche,* edited by J. Fletcher, 260–65. London: Routledge, 1999.

Lardas, John. *The Bop Apocalypse: The Religious Visions of Kerouac, Ginsberg, and Burroughs.* Urbana and Chicago: University of Illinois Press, 2001.

Larson, Erik. *The Devil in the White City: Murder, Magic, Madness, and the Fair that Changed America.* New York: Vintage, 2003.

Lasch, Christopher. *The Culture of Narcissism: American Life in an Age of Diminishing Expectations.* New York and London: W. W. Norton & Company, 1979.

Laurence, William L. *Dawn over Zero: The Story of the Atomic Bomb.* Westport, CT: Greenwood Press, 1977. First published 1946.

Lee, Ben. "'Howl' and Other Poems: Is There Old Left in These New Beats?" *American Literature,* 76:2 (June 2004): 367–89.

LeMenager, Stephanie. "The Aesthetics of Petroleum, after *Oil!*" *American Literary History,* 24:1 (Spring 2012): 59–86.

———. *Living Oil: Petroleum Culture in the American Century.* Oxford and New York: Oxford University Press, 2014. Ebook.

Lentricchia, Frank, and Jody McAuliffe. *Crimes of Art + Terror.* Chicago and London: University of Chicago Press, 2003.

Les Doigts de la Main. "Traces." http://7doigts.com/en/shows/6-traces.

Libeskind, Daniel. *Breaking Ground: Adventures in Life and Architecture.* New York: Riverhead Books, 2004.

Lifset, Robert, and Brian C. Black. "Imaging the 'Devil's Excrement': Big Oil in Petroleum Cinema, 1940–2007." *Journal of American History,* 99:1 (June 2012): 135–44.

Lincoln, Kenneth. *Native American Renaissance.* Berkeley: University of California Press, 1983.

Lobo, Susan, and Kurt Peters. *American Indians and the Urban Experience.* Walnut Creek, CA, and Oxford: Altimira Press, 2001.

Longmuir, Anne. "'This Was the World Now': *Falling Man* and the Role of the Artist after 9/11." *Modern Language Studies,* 41:1 (Summer 2011): 42–57.

Lovejoy, Arthur O. *The Great Chain of Being: A Study of the History of an Idea.* Cambridge, MA: Harvard University Press, 1961. First published 1936.

Lurie, Susan. "Falling Persons and National Embodiment: The Reconstruction of Safe Spectatorship in the Photographic Record of 9/11." In *Terror, Culture, Politics: Rethinking 9/11,* edited by Daniel J. Sherman and Terry Nardin, 44–68. Bloomington and Indianapolis: Indiana University Press, 2006.

———. "Spectacular Bodies and Political Knowledge: 9/11 Cultures and the Problem of Dissent." *American Literary History,* 25:1 (Spring 2013): 176–89.

Mackay, Ruth. "'Going Backwards in Time to Talk about the Present': *Man on Wire* and Verticality after 9/11." *Comparative American Studies,* 9:1 (March 2011): 3–20.

———. "Representing 9/11: The Kinetic Axes of Direction in Jonathan Safran Foer's *Extremely Loud and Incredibly Close* and Art Spiegelman's *In the Shadow of No Towers.*" MA thesis, School of English, University of Leeds, 2008.

Mackay, Ruth, and Stephen Mitchell. "Re-constructing 'Le Coup': *Man on Wire*, Derrida's Event, and Cinematic Representation." In *Dramatising Disaster: Character, Event, Representation,* edited by Christine Cornea and Rhys Owain Thomas, 58–71. Newcastle upon Tyne, UK: Cambridge Scholars Publishing, 2013.

MacKendrick, Karmen. "Embodying Transgression: Transgressive Intensification." In *Of the Presence of the Body: Essays on Dance and Performance Theory,* edited by Andre Lepecki, 140–56. Middletown, CT: Wesleyan University Press, 2004.

Makhmalbaf, Samira, Claude Lelouch, Youssef Chahine, Danis Tanovic, Idrissa Ouedraogo, Ken Loach, Alejandro Gonzalez Inarritu, Amos Gitai, Mira Nair, Sean Penn, and Shohei Imamura. *11'09"01: September 11,* DVD. World Cinema, Ltd., 2002.

Marcus, Laura. "Cinema and Psychoanalysis." In *Close Up 1927–1933: Cinema and Modernism,* edited by James Donald, Anne Friedberg, and Laura Marcus, 240–46. London: Cassell, 1998.

———. "Introduction: The Contribution of H. D." Introduction to *Close Up 1927–1933: Cinema and Modernism,* edited by James Donald, Anne Friedberg and Laura Marcus, 96–104. London: Cassell, 1998.

———. "'A New Form of True Beauty': Aesthetics and Early Film Criticism." *MODERNISM/modernity,* 13:2 (April 2006): 267–89.

———. *The Tenth Muse: Writing about Cinema in the Modernist Period.* Oxford: Oxford University Press, 2007.

Marschall, Richard, ed. *The Complete Little Nemo in Slumberland, by Winsor McCay, Volume II: 1907–1908.* London: Titan Books, 1989.

Marsh, James. *Man on Wire,* DVD. Icon, 2008.

Marx, Doug. "Sherman Alexie: A Reservation of the Mind." *Publishers Weekly,* 16 September 1996, 39–40.

Marx, Leo. *The Machine in the Garden.* Oxford and New York: Oxford University Press, 2000. First published 1964.

Mather, Ronald, and Jill Marsden. "Trauma and Temporality: On the Origins of Post-Traumatic Stress." *Theory and Psychology,* 14:2 (2004): 205–19.

Mathews, John Joseph. *Sundown.* Norman: University of Oklahoma Press, 1987. First published 1934.

———. *Talking to the Moon.* Norman: University of Oklahoma Press, 1981. First published 1945.

Mauro, Aaron. "The Languishing of the Falling Man: Don DeLillo and Jonathan Safran Foer's Photographic History of 9/11." *Modern Fiction Studies,* 57:3 (Fall 2011): 584–606.

Mavroudis, John, and Owen Smith. "Soaring Spirit," front and inside cover of *The New Yorker,* 11 September 2006.

McCann, Colum. *Let the Great World Spin.* London, Berlin, and New York: Bloomsbury, 2009.

McCay, Winsor. *Dream of the Rarebit Fiend.* New York: Dover Publications, 1974.

———. *Little Nemo: Little Nemo in Slumberland and Little Nemo in the Land of Wonderful Dreams 1905–1914.* China: Evergreen, 2000.

McCloud, Scott. *Understanding Comics: The Invisible Art.* New York: HarperPerennial, 1994.

McLaurin, John J. *Sketches in Crude-Oil: Some Accidents and Incidents of the Petroleum Development in All Parts of the Globe.* Franklin, PA: John J. McLaurin, 1902. First published 1896.

McNickle, D'Arcy. *The Surrounded.* Albuquerque: University of New Mexico Press, 2003. First published 1936.

Melville, Herman. *Moby-Dick, or the Whale.* Evanston, IL: Northwestern University Press and The Newberry Library, 1988. First published 1851.

Merriam-Webster's Intermediate Dictionary. Springfield, MA: Merriam-Webster, 2004.

Meyerowitz, Joel. *Aftermath: World Trade Center Archive.* New York: Phaidon, 2006.

Micale, Mark S., ed. *The Mind of Modernism: Medicine, Psychology, and the Cultural Arts in Europe and America, 1880–1940.* Stanford, CA: Stanford University Press, 2004.

Micale, Mark S. "The Modernist Mind: A Map." In *The Mind of Modernism: Medicine, Psychology, and the Cultural Arts in Europe and America, 1880–1940,* 1–19. Stanford, CA: Stanford University Press, 2004.

Mickalites, Carey James. "Manhattan Transfer, Spectacular Time and the Outmoded." *Arizona Quarterly,* 67:4 (Winter 2011): 59–82.

Miles, Barry. *Allen Ginsberg: A Biography.* London: Virgin Books, 2002. First published 1989.

Miller, Nancy. "'Portraits of Grief': Telling Details and the Testimony of Trauma." *Differences: A Journal of Feminist Cultural Studies,* 14:3 (2003): 112–35.

Mitchell, William J. T. *What Do Pictures Want? The Lives and Loves of Images.* Chicago and London: University of Chicago Press, 2005.

Momaday, N. Scott. *House Made of Dawn.* New Delhi: Asian Books, 1968.

———. *The Way to Rainy Mountain.* Albuquerque: University of New Mexico, 1969.

Mookerjee, R. N. *Art for Social Justice: The Major Novels of Upton Sinclair.* Metuchen, NJ: Scarecrow Press, 1988.

Moore, Rowan. "Ground Zero 9/11 Memorial Flows with Mournful Splendour." *The Guardian,* 15 August 2011. http://www.guardian.co.uk/artanddesign/2011/aug/15/ground-zero-memorial-september-9-11?INTCMP=SRCH.

Morley, Catherine. "'How Do We Write about This?' The Domestic and the Global in the Post-9/11 Novel." *Journal of American Studies,* 45:4 (November 2011): 717–31.

———. "Writing in the Wake of 9/11." In *American Thought and Culture in the 21st Century,* edited by Martin Halliwell and Catherine Morley, 245–58. Edinburgh: Edinburgh University Press, 2008.

Morritt, Hope. *Rivers of Oil: The Founding of North America's Petroleum Industry.* Toronto, ON: Quarry Press, 1993.

Morton, Patricia A. "'Document of Civilization and Document of Barbarism': The World Trade Center Near and Far." In *Terror, Culture, Politics: Rethinking 9/11,* edited by Daniel J. Sherman and Terry Nardin, 15–32. Bloomington and Indianapolis: Indiana University Press, 2006.

Mott, Frank Luther. *American Journalism: 1690–1940,* Vol. 5. London: Routledge, 2000.

Moudry, Roberta, ed. *The American Skyscraper: Cultural Histories*. Cambridge: Cambridge University Press, 2005.

Moynihan, Sinéad. "'Upground and Belowground Topographies': The Chronotypes of Skyscraper and Subway in Colum McCann's New York Novels before and after 9/11." *Studies in American Fiction*, 39:2 (2012): 269–90.

Mumford, Lewis. *The Brown Decades: A Study of the Arts in America 1865–1895*. New York: Dover Publications, 1971. First published 1931.

———. *The City in History: Its Origins, Its Transformations, and Its Prospects*. Harmondsworth, UK: Penguin, 1991.

Muntean, Nick. "'It Was Just Like a Movie': Trauma, Memory, and the Mediation of 9/11." *Journal of Popular Film and Television*, 37:2 (Summer 2009): 50–58.

Murphy, Emily. "Life on the Wire: Post-9/11 Mourning in Mordicai Gerstein's *The Man Who Walked between the Towers*." *The Lion and the Unicorn*, 38:1 (2014): 66–85.

Nadel, Ira. "White Rain: 9/11 and American Fiction." *Canadian Review of American Studies*, 45:2 (Summer 2015): 125–48.

Nash, Robert Frazier. *Wilderness and the American Mind*, 4th ed. New Haven, CT, and London: Yale University Press, 2001. First published 1967.

Naudet, Jules, Gedeon Naudet, and James Hanlon. *9/11*, DVD. Goldfish Pictures, 2001.

Negra, Diana. "Structural Integrity, Historical Reversion, and the Post-9/11 Chick Flick." *Feminist Media Studies*, 8:1 (2008): 51–68.

Nelson, Joshua B. "Fight as Flight: The Traditional Reclamation of Exploration." *World Literature Today*, 84:4 (2010): 22–47.

New York Times. "Attacks Called Great Art." 19 September 2001. http://www.nytimes.com/2001/09/19/arts/attacks-called-great-art.html

Niezen, Ronald. *The Origins of Indigenism: Human Rights and the Politics of Identity*. Berkeley, Los Angeles, and London: University of California Press, 2003.

Nobel, Philip. *Sixteen Acres: Architecture and the Outrageous Struggle for the Future of Ground Zero*. New York: Henry Holt & Company, 2005.

Nye, David E. "Accelerating American Experience: The Values Embodied in Assembly Line." *Odense American Studies International Series*, 68 (2005).

———. *America as Second Creation: Technology and Narratives of New Beginnings*. Cambridge, MA, and London: MIT Press, 2003.

———. *American Technological Sublime*. Cambridge, MA, and London: MIT Press, 1999.

———. *Electrifying America: Social Meanings of a New Technology, 1880–1940*. Cambridge, MA: MIT Press, 1990.

———. *Narratives and Spaces: Technology and the Construction of American Culture*. New York: Columbia University Press, 1997.

———. "The Sublime and the Skyline." In *The American Skyscraper: Cultural Histories*, edited by Roberta Moudry, 255–69. Cambridge: Cambridge University Press, 2005.

Nye, David E., and Thomas Johansen. *The American Century: A Chronology and Orientation*. Odense: University Press of Southern Denmark, 2007.

O'Henry, John. "Psyche and the Pskyscraper." http://www.literaturecollection.com/a/o_henry/197/

O'Meara, Bridget. "The Ecological Politics of Leslie Marmon Silko's *Almanac of the Dead.*" *Wicazo Sa Review,* 15:2 (Fall 2000): 63–73.

Olmsted, Jane. "The Uses of Blood in Leslie Marmon Silko's *Almanac of the Dead.*" *Contemporary Literature,* 40:3 (Autumn 1999): 464–90.

Olster, Stacey, ed. *Don DeLillo: Mao II, Underworld, Falling Man.* London and New York: Continuum, 2011.

Op de Beeck, Nathalie. "Found Objects: (Jem Cohen, Ben Katchor, Walter Benjamin)." *Modern Fiction Studies,* 52:4 (Winter 2006): 807–30.

Orbán, Katalin. "Trauma and Visuality: Art Spiegelman's *Maus* and *In the Shadow of No Towers.*" *Representations,* 97:1 (Winter 2007): 57–89.

Ortiz, Alfonso. "Indian/White Relations: A View from the Other Side of the 'Frontier.'" In *Indians in American History,* edited by Frederick E. Hoxie and Peter Iverson, 1–14. Wheeling, IL: Harlan Davidson, 1998.

Owens, Louis. *Mixedblood Messages: Literature, Film, Family, Place.* Norman: University of Oklahoma Press, 1998.

———. *Other Destinies: Understanding the American Indian Novel.* Norman: University of Oklahoma Press, 1992.

Oxley, Chris. *Let's Roll: The Story of Flight 93.* Granada Television, 2002.

Page, Max. *The City's End: Two Centuries of Fantasies, Fears, and Premonitions of New York's Destruction.* New Haven, CT, and London: Yale University Press, 2008.

———. *The Creative Destruction of Manhattan, 1900–1940.* Chicago and London: University of Chicago Press, 1999.

———. "The Heights and Depths of Urbanism: Fifth Avenue and the Creative Destruction of Manhattan." In *The American Skyscraper: Cultural Histories,* edited by Roberta Moudry, 165–84. Cambridge: Cambridge University Press, 2005.

Pease, Donald E. *The New American Exceptionalism.* Minneapolis and London: University of Minnesota Press, 2009.

Petit, Philippe. *To Reach the Clouds: My High Wire Walk between the Twin Towers.* London: Faber and Faber, 2003. First published 2002.

Picturing a Metropolis: New York City Unveiled, DVD. Image Entertainment, 2005.

Pike, David L. "'Kaliko-Welt': The Großstädte of Lang's *Metropolis* and Brecht's *Dreigroschenoper.*" *MLN,* 119:3 (April 2004): 474–505.

———. *Metropolis on the Styx: The Underworlds of Modern Urban Culture, 1800–2001.* Ithaca, NY, and London: Cornell University Press, 2007.

———. *Subterranean Cities: The World beneath Paris and London, 1800–1945.* Ithaca, NY, and London: Cornell University Press, 2005.

———. "Urban Nightmares and Future Visions: Life beneath New York." *Wide Angle,* 20:4 (October 1998): 8–50.

Pilkington, Ed. "9/11 Ten Years On: America's Tallest Building Rises from the Rubble of Ground Zero." *The Guardian,* 4 September 2011. http://www.guardian.co.uk/world/2011/sep/04/911-ground-zero-skyscraper.

Pinsker, Sanford. "Henry Adams at Ground Zero." *Virginia Quarterly Review,* 78:2 (Spring 2002): 189–99. http://www.vqronline.org/essay/henry-adams-ground-zero.

Pomerance, Murray. "The Shadow of the World Trade Center Is Climbing My Memory of Civilization." In *Film and Television after 9/11*, edited by Wheeler Winston Dixon, 42–62. Carbondale: Southern Illinois University Press, 2004.

Porter, Joy. "Imagining Indians: Differing Perspectives on Native American History." In *The State of U.S. History*, edited by Melvyn Stokes, 347–66. Oxford and New York: Berg, 2002.

———. "Historical and Cultural Contexts to Native American Literature." In *Cambridge Companion to Native American Literature*, 39–68. Cambridge: Cambridge University Press, 2005.

Pozorski, Aimee. *Falling after 9/11: Crisis in American Art and Literature*. New York and London: Bloomsbury, 2014.

Pratt, Mary Louise. *Imperial Eyes: Travel Writing and Transculturation*. London: Routledge, 2008. First published 1992.

Prucha, Francis Paul. *The Great Father: The United States Government and the American Indians*. Lincoln: University of Nebraska Press, 1984.

Quam-Wickham, Nancy. "'Cities Sacrificed on the Altar of Oil': Popular Opposition to Oil Development in 1920s Los Angeles." *Environmental History*, 3:2 (April 1998): 189–209.

Quinn, Justin. "Coteries, Landscape and the Sublime in Allen Ginsberg." *Journal of Modern Literature*, 27:1/2 (Fall 2003): 193–206.

Raeburn, Daniel. *Chris Ware*. London: Laurence King Publishing, 2004.

Randall, Martin. *9/11 and the Literature of Terror*. Edinburgh: Edinburgh University Press, 2011.

———. "'A Certain Blurring of the Facts': *Man on Wire* and 9/11." In *9/11 and the Literature of Terror*, 88–98. Edinburgh: Edinburgh University Press, 2011.

Raskin, Jonah. *American Scream: Allen Ginsberg's* Howl *and the Making of the Beat Generation*. Berkeley, Los Angeles, and London: University of California Press, 2005.

Reed, T. V. "Toxic Colonialism, Environmental Justice, and Native Resistance in Silko's *Almanac of the Dead*." *MELUS*, 34:2 (Summer 2009): 25–42.

Richter, Daniel K. *Facing East from Indian Country: A Native History of Early America*. Boston: Harvard University Press, 2003.

Riese, Utz. "Upton Sinclair's Contribution to a Proletarian Aesthetic." In *Upton Sinclair: Literature and Social Reform*, edited by Dieter Herms, 11–23. Frankfurt am Main, Bern, New York, and Paris: Peter Lang, 1990.

Riffel, Casey. "Dissecting Bambi: Multiplanar Photography, the Cel Technique, and the Flowering of Full Animation." *Velvet Light Trap*, 69 (Spring 2012): 3–16.

Rifkin, Mark. *Manifesting America: The Imperial Construction of U.S. National Space*. Oxford: Oxford University Press, 2009.

Robb, Jenny E. "Winsor McCay, George Randolph Chester, and the Tale of *The Jungle Imps*." *American Periodicals*, 17:2 (2007): 245–59.

Roeder, Katherine. "Seeing Inside-out in the Funny Pages." *American Art*, 25:1 (Spring 2011): 24–27.

———. *Wide Awake in Slumberland: Fantasy, Mass Culture, and Modernism in the Art of Winsor McCay*. Jackson: University Press of Mississippi, 2014.

Romero, Channette. "Envisioning A 'Network of Tribal Coalitions': Leslie Marmon Silko's *Almanac of the Dead*." *American Indian Quarterly*, 26:4 (Autumn 2002): 623–40.

Rosenthal, Peggy. "The Nuclear Mushroom Cloud as Cultural Image." *American Literary History*, 3:1 (Spring 1991): 63–92.

Rosenzweig, Saul. *Freud, Jung and Hall the King-Maker: The Historic Expedition to America*. St. Louis, MO: Rana House Press, 1992. First published 1909.

Ross, Luana. *Inventing the Savage: The Social Construction of Native American Criminality*. Austin: University of Texas Press, 1998.

Roszak, Theodore. *The Making of a Counterculture: Reflections on the Technocratic Society and Its Youthful Opposition*. Berkeley, Los Angeles, and London: University of California Press, 1995. First published 1968.

Rothberg, Michael. Preface to *The Future of Trauma Theory: Contemporary Literary and Cultural Criticism*, edited by Gert Buelens, Sam Durrant, and Robert Eaglestone, xi–xvii. London and New York: Routledge, 2014.

Rowe, John Carlos. "Culture, U.S. Imperialism, and Globalization." In *Exceptional State: Contemporary U.S. Culture and the New Imperialism*, edited by Ashley Dawson and Malini Johar Schueller, 37–59. Durham, NC, and London: Duke University Press, 2007.

———. "Global Horizons in *Falling Man*." In *Don DeLillo: Mao II, Underworld, Falling Man*, edited by Stacey Olster, 121–34. London and New York: Continuum, 2011.

———. Introduction to *New Essays on the Education of Henry Adams*, edited by John Carlos Rowe, 1–20. Cambridge: Cambridge University Press, 1996.

Ryan, Judith. *The Vanishing Subject: Early Psychology and Literary Modernism*. Chicago: University of Chicago Press, 1991.

Saal, Ilka. "Regarding the Pain of Self and Other: Trauma Transfer and Narrative Framing in Jonathan Safran Foer's *Extremely Loud and Incredibly Close*." *Modern Fiction Studies*, 57:3 (Fall 2011): 453–76.

Saalschutz, L. "The Film in Its Relation to the Unconscious." Originally published in *Close Up*, 5:1 (July 1929), reprinted in *Close Up 1927–1933: Cinema and Modernism*, edited by James Donald, Anne Friedberg, and Laura Marcus, 256–60. London: Cassell, 1998.

Sabin, Paul. *Crude Politics: The California Oil Market, 1900–1940*. Berkeley and London: University of California Press, 2005.

Sabin, Roger. *Comics, Comix and Graphic Novels: A History of Comic Art*. London: Phaidon, 1996.

Salaita, Steven. "Concocting Terrorism off the Reservation: Liberal Orientalism in Sherman Alexie's Post-9/11 Fiction." *Studies in American Indian Literatures*, 22:2 (Summer 2010): 22–41.

Salvidar, Jose David. *Remapping American Cultural Studies*. Berkeley, London, and Los Angeles: University of California Press, 1997.

Sánchez-Escalonilla, Antonio. "The Hero as a Visitor in Hell: The Descent into Death in Film Structure." *Journal of Popular Film and Television*, 32:4 (Winter 2005): 149–56.

———. "Hollywood and the Rhetoric of Panic: The Popular Genres of Action and Fantasy in the Wake of the 9/11 Attacks." *Journal of Popular Film and Television*, 38:1 (January–March 2010): 10–20.

Sattler, Peter R. "Past Imperfect: 'Building Stories' and the Art of Memory." In *The Comics of Chris Ware: Drawing Is a Way of Thinking*, edited by David M. Ball and Martha B. Kuhlman, 206–22. Jackson: University Press of Mississippi, 2010.

Savage, Kirk. "Trauma, Healing, and the Therapeutic Monument." In *Terror, Culture, Politics: Rethinking 9/11*, edited by Daniel J. Sherman and Terry Nardin, 103–20. Bloomington and Indianapolis: Indiana University Press, 2006.

Schleier, Merrill. *Skyscraper Cinema: Architecture and Gender in American Film*. Minneapolis and London: University of Minnesota Press, 2009.

———. *The Skyscraper in American Art, 1890–1931*. New York: Da Capo, 1986.

Schneider, Steven Jay. "Architectural Nostalgia and the New York City Skyline on Film." In *Film and Television after 9/11*, edited by Wheeler Winston Dixon, 29–41. Carbondale: Southern Illinois University Press, 2004.

Scobey, David M. *Empire City: The Making and Meaning of the New York City Landscape*. Philadelphia: Temple University Press, 2002.

Scott, Ivan. *Upton Sinclair, the Forgotten Socialist, Studies in American Literature*. Lewiston, NY: Edwin Mellen Press, 1997.

Scraton, Phil, ed. *Beyond September 11: An Anthology of Dissent*. London and Sterling, VA: Pluto Press, 2002.

Senk, Sarah. "Lost in Space." *American Prospect*, 12 September 2011. http://prospect.org/article/lost-space.

Shannon, Edward A. "Review of Little Nemo in Slumberland: Splendid Sundays by Winsor McCay, edited by Peter Maresca." *Modern Language Studies*, 37:1 (Summer 2007): 93–98.

———. "Something Black in the American Psyche: Formal Innovation and Freudian Imagery in the Comics of Winsor McCay and Robert Crumb." *Canadian Review of American Studies*, 40:2 (July 2010): 187–211.

Shannon, Laurie, Vin Nardizzi, Ken Hiltner, Saree Makdisi, Michael Ziser, and Imre Szeman. "Editor's Column: Literature in the Ages of Wood, Tallow, Coal, Whale Oil, Gasoline, Atomic Power, and Other Energy Sources." *PMLA*, 126:2 (March 2011): 305–26.

Sheehan, Bernard W. *Seeds of Extinction: Jeffersonian Philanthropy and the American Indian*. New York: W. W. Norton & Company, 1973.

Sherman, Daniel J. "Naming and the Violence of Place." In *Terror, Culture, Politics: Rethinking 9/11*, edited by Daniel J. Sherman and Terry Nardin, 121–45. Bloomington and Indianapolis: Indiana University Press, 2006.

Sherman, Daniel J., and Terry Nardin, eds. *Terror, Culture, Politics: Rethinking 9/11*. Bloomington and Indianapolis: Indiana University Press, 2006.

Silko, Leslie Marmon. *Almanac of the Dead*. London: Penguin, 1992.

———. "The Border Patrol State." In *Yellow Women and a Beauty of the Spirit: Essays on Native American Life Today*, 115–23. New York: Simon and Schuster, 1996.

———. *Ceremony*. London: Penguin, 1977.

———. "Fences against Freedom." In *Yellow Women and a Beauty of the Spirit: Essays on Native American Life Today*, 100–14. New York: Simon and Schuster, 1996.

———. "Notes on *Almanac of the Dead*." In *Yellow Women and a Beauty of the Spirit: Essays on Native American Life Today*, 135–45. New York: Simon and Schuster, 1996.

———. "The People and the Land Are Inseparable." In *Yellow Women and a Beauty of the Spirit: Essays on Native American Life Today,* 85–91. New York: Simon and Schuster, 1996.

———. *A Range of Poems.* London: Fulcrum Press, 1967.

———. *Yellow Women and a Beauty of the Spirit: Essays on Native American Life Today.* New York: Simon and Schuster, 1996.

Simpson, David. *9/11: The Culture of Commemoration.* Chicago and London: University of Chicago Press, 2006.

———. "Shortcuts." *London Review of Books,* 17 November 2011, 22.

Sinclair, Upton. *The Autobiography of Upton Sinclair.* New York: Harcourt, Brace & World, Inc., 1962.

———. *Candid Reminiscences: My First Thirty Years.* London: T. Werner Laurie, Ltd., 1932.

———. *The Jungle.* New York: Doubleday, Page & Company, 1906.

———. *The Millennium: A Comedy of the Year 2000.* London: T. Werner Laurie, Ltd., 1929. First published 1924.

———. *Oil!* London: Penguin, 2008. First published 1927.

Slotkin, Richard. *Gunfighter Nation: The Myth of the Frontier in Twentieth-Century America.* New York and Oxford: Maxwell Macmillan, 1992.

———. *Regeneration through Violence: The Mythology of the American Frontier, 1600–1860.* Norman: University of Oklahoma Press, 2000. First published 1973.

Smith, Henry Nash. *Virgin Land: The American West as Symbol and Myth.* Cambridge, MA: Harvard University Press, 1970. First published 1950.

Smith, Neil. "Wire-Walk Film Omits 9/11 Tragedy." BBC News, 2 August 2008. http://news.bbc.co.uk/1/hi/entertainment/7498364.stm.

Smolderen, Thierry. *The Origins of Comics: From William Hogarth to Winsor McCay.* Translated by Bart Beaty and Nick Nguyen. Jackson: University Press of Mississippi, 2000.

Snyder, Gary. *A Place in Space: Ethics, Aesthetics, and Watersheds.* New York: Counterpoint, 1995.

———. *A Range of Poems.* London: Fulcrum Press, 1967.

Sontag, Susan. *Regarding the Pain of Others.* London: Penguin 2004. First published 2003.

———. "Regarding the Torture of Others." *New York Times,* 23 May 2004. http://www.nytimes.com/2004/05/23/magazine/regarding-the-torture-of-others.html?pagewanted=all&src=pm

Spiegelman, Art. *Breakdowns: Portrait of the Artist as a Young %@ &*!.* London: Penguin, 2008.

———. *The Complete Maus.* London: Penguin, 2003. First printed 1986, again in 1992.

———. *In the Shadow of No Towers.* New York: Random House, 2004.

———. "The Sky Is Falling!" In *In the Shadow of No Towers,* i–ii. New York: Random House, 2004.

Stauffer, John. Foreword to *American Protest Literature,* edited by Zoe Trodd, xi–xviii. Cambridge, MA, and London: Harvard University Press, 2006.

Stone, Oliver. *World Trade Center*, DVD. Paramount, 2006.

Sturken, Marita. "The 9/11 Memorial Museum and the Remaking of Ground Zero." *American Quarterly*, 67:2 (June 2015): 471–90.

———. "Memorializing Absence." http://essays.ssrc.org/sept11/essays/sturken.htm.

———. *Tourists of History: Memory, Kitsch, and Consumerism from Oklahoma City to Ground Zero*. Durham, NC, and London: Duke University Press, 2007.

Suarez, Juan A. "City Films, Modern Spatiality, and the End of the World Trade Center." In *Film and Television after 9/11*, edited by Wheeler Winston Dixon, 101–20. Cardondale: Southern Illinois University Press, 2004.

———. "City Space, Technology, Popular Culture: The Modernism of Paul Strand and Charles Sheeler's *Manhatta*." *Journal of American Studies*, 36:1 (April 2002): 85–106.

Sullivan, Louis H., "The Tall Office Building Artistically Considered." *Lippincott's Magazine*, 57, March 1896, 403–9.

Svonkin, Craig. "Manishevitz and Sake, the Kaddish and Sutras: Allen Ginsberg's Spiritual Self-Othering." *College Literature*, 37:4 (Fall 2010): 166–93.

Swann, Brian, and Arnold Krupat, eds. *Recovering the Word: Essays on Native American Literature*. Berkeley, Los Angeles, and London: University of California Press, 1987.

Swartz, Anne K. "American Art after 9/11: A Consideration of the Twin Towers." *symploke*, 14:1–2 (2006): 81–97.

Tallmadge, Thomas E. *The Story of Architecture in America*. New York: W. W. Norton & Company, 1936.

Taylor, Bryan C. "Nuclear Pictures and Metapictures." *American Literary History*, 9:3 (Autumn 1997): 567–97.

Tebbel, John. *The Compact History of the American Newspaper*. New York: Hawthorn Books, 1963.

Thrapp, Dan. *The Conquest of Apacheria*. Norman: University of Oklahoma University, 1967.

Tillett, Rebecca. "The Price of 'Free' Trade: NAFTA and the Economies of Border Crossing in George Rabasa's *The Floating Kingdom* and Leslie Marmon Silko's *Almanac of the Dead*." *Comparative American Studies*, 6:4 (December 2008): 329–43.

Trachtenberg, Alan. *The Incorporation of America: Culture and Society in the Gilded Age*. New York: Hill and Wang, 1982.

———. *Reading American Photographs: Images as History from Mathew Brady to Walker Evans*. New York: Hill and Wang, 1989.

Trafzer, Clifford E. *As Long as the Grass Shall Grow and Rivers Flow: A History of Native Americans*. Riverside: University of California, 2000.

Trodd, Zoe, ed. *American Protest Literature*. Cambridge, MA, and London: Harvard University Press, 2006.

Turner, Frederick Jackson. *The Frontier in American History*. New York: Holt, Rinehart, and Winston, 1962. First published 1920.

Upton, Florence Kate. *Two Dutch Dolls and a Golliwogg*. London: Dodo Press, 2007. First published 1895.

Vanderwees, Chris. "Photographs of Falling Bodies and the Ethics of Vulnerability in Jonathan Safran Foer's *Extremely Loud and Incredibly Close*." *Canadian Review of American Studies*, 45:2 (Summer 2015): 174–94.

———. "A Tightrope at the Twin Towers: Falling Bodies and James Marsh's *Man on Wire*." In *Recovering 9/11 in New York*, edited by Robert Fanuzzi and Michael Wolfe, 228–47. Newcastle upon Tyne, UK: Cambridge Scholars, 2014.

Van Leeuwen, Thomas A. P. *The Skyward Trend of Thought: The Metaphysics of the American Skyscraper*. Cambridge, MA: MIT Press, 1988.

Van Wyck, Peter C. "American Monument: The Waste Isolation Pilot Plant." In *Atomic Culture: How We Learned to Stop Worrying and Love the Bomb*, edited by Scott C. Zeman and Michael A. Amundson, 149–72. Boulder, CO: University Press of Colorado, 2004.

Vargish, Thomas, and Delo E. Mook. *Inside Modernism: Relativity Theory, Cubism, Narrative*. New Delhi: Yale University Press, 1999.

Varnum, Robin, and Christina T. Gibbons, eds. *The Language of Comics: Word and Image*. Jackson: University Press of Mississippi, 2001.

Vermeulen, Pieter. "The Biopolitics of Trauma." In *The Future of Trauma Theory: Contemporary Literary and Cultural Criticism*, edited by Gert Buelens, Sam Durrant, and Robert Eaglestone, 141–55. London and New York: Routledge, 2014.

Versluys, Kristiaan. "Art Spiegelman's *In the Shadow of No Towers*: 9/11 and the Representation of Trauma." *Modern Fiction Studies*, 52:4 (Winter 2006): 980–1003.

———. *Out of the Blue: September 11 and the Novel*. New York: Columbia University Press, 2009.

Virilio, Paul. *Ground Zero*. Translated by Chris Turner. London and New York: Verso, 2002.

Vizenor, Gerald. "Aesthetics of Survivance: Literary Theory and Practice." In *Survivance: Narratives of Native Presence*, edited by Gerald Vizenor, 1–24. Lincoln: University of Nebraska Press, 2008.

———. *Fugitive Poses: Native American Indian Scenes of Absence and Presence*. Lincoln and London: University of Nebraska Press, 1998.

———. *Manifest Manners: Postindian Warriors of Survivance*. Hanover and London: Wesleyan University Press, 1994.

———, ed. *Narrative Chance: Postmodern Discourse on Native American Indian Literatures*. Norman: University of Oklahoma Press, 1993.

———, ed. *Native American Literature: A Brief Introduction and Anthology*. New York and Harlow, UK: Longman, 1995.

———. *Native Liberty: Natural Reason and Cultural Survivance*. Lincoln: University of Nebraska Press, 2009.

———, ed. *Survivance: Narratives of Native Presence*. Lincoln: University of Nebraska Press, 2008.

Vizenor, Gerald, and A. Robert Lee. *Postindian Conversations*. Lincoln: University of Nebraska Press, 1999.

Von Dassanowsky, Robert. "A Caper of One's Own: Fantasy Female Liberation in the 1960s Crime Comedy Film." *Journal of Popular Film and Television*, 35:3 (Fall 2007): 107–18.

Vulliamy, Ed. *Amexica: War along the Borderline*. New York: Farrar, Straus, and Giroux, 2010.

Walker, Cheryl. *Indian Nation: Native American Literature and Nineteenth-Century Nationalisms*. Durham, NC, and London: Duke University Press, 1997.

Wallace, Anthony F. C. *Jefferson and the Indians: The Tragic Fate of the First Americans.* Cambridge, MA, and London: Harvard University Press, 1999.

Walonen, Michael K. "'The Black and Cruel Demon' and Its Transformations of Space: Toward a Comparative Study of the World Literature of Oil and Place." *Interdisciplinary Literary Studies,* 14:1 (2012): 56–78.

Ward, Paul. "Drama-Documentary, Ethics and Notions of Performance: The 'Flight 93' Films." In *Rethinking Documentary: New Perspectives, New Practices,* edited by Thomas Austin and Wilma de Jong, 191–203. Maidenhead, Berkshire: Open University Press, 2008.

Ware, Chris. *Building Stories.* New York: Pantheon, 2012.

———. *Jimmy Corrigan: The Smartest Kid on Earth.* London: Jonathan Cape, 2003.

Waters, Frank. *Book of the Hopi.* New York: Random House, 1963.

Weaver, Jace. "The Mystery of Language: N. Scott Momaday, an Appreciation." *Studies in American Indian Literatures,* 20:4 (Winter 2008): 76–86.

———. "Native American Authors and Their Communities." *Wicazo Sa Review,* 12:1 (Spring 1997): 47–87.

———. *That the People Might Live: Native American Literatures and Native American Community.* New York: Oxford University Press, 1997.

Weaver, Jace, Craig S. Womack, and Robert Warrior, eds. *American Indian Literary Nationalism.* Alberquerque: University of New Mexico Press, 2006.

Webb, Walter Prescott. "The American Frontier Concept." In *The Frontier in American Literature,* edited by Philip Durham and Everett L. Jones, 3–8. New York: Odyssey Press, 1969. First published 1952.

West, Elliot. *The Contested Plains: Indians, Goldseekers, and the Rush to Colorado.* Lawrence: University Press of Kansas, 1998.

Whissel, Kristen. *Spectacular Digital Effects: CGI and Contemporary Cinema.* Durham, NC, and London: Duke University Press, 2014.

———. "Tales of Upward Mobility." *Film Quarterly,* 59:4 (Summer 2006): 23–34.

Wilson, Rob. *American Sublime: The Genealogy of a Poetic Genre.* Madison: University of Wisconsin Press, 1991.

Womack, Craig S. *Drowning in Fire.* Tucson: University of Arizona Press, 2001.

———. *Red on Red: Native American Literary Separatism.* Minneapolis: University of Minnesota Press, 1999.

Woolf, Virginia. "Modern Fiction." In *The Common Reader,* 184–95. London: Hogarth Press, 1925.

Worden, Daniel. "On Modernism's Ruins: The Architecture of "Building Stories" and Lost Buildings." In *The Comics of Chris Ware: Drawing Is a Way of Thinking,* edited by David M. Ball and Martha B. Kuhlman, 107–20. Jackson: University Press of Mississippi, 2010.

Wyatt, Jean. "*Love*'s Time and the Reader: Ethical Effects of Nachträglichkeit in Toni Morrison's *Love.*" *Narrative,* 16:2 (May 2008): 193–221.

Yablon, Nick. "The Metropolitan Life in Ruins: Architectural and Fictional Speculations in New York, 1909–19." *American Quarterly,* 56:2 (June 2004): 309–47.

———. *Untimely Ruins: An Archaeology of American Urban Modernity, 1819–1919.* Chicago: Chicago University Press, 2010.

Yoder, Jon A. *Upton Sinclair.* New York: Frederick Ungar Publishing Company, 1975.

Zeman, Scott C., and Michael A. Amundson, eds. *Atomic Culture: How We Learned to Stop Worrying and Love the Bomb.* Boulder, CO: University Press of Colorado, 2004.

Zitkala-Sa. "The School Days of an Indian Girl." *Atlantic Monthly,* 85:508 (February 1900): 185–94.

Žižek, Slavoj. *Welcome to the Desert of the Real!* London: Verso, 2002.

INDEX

11 September 2001. *See* 9/11
9/11 (Naudet, Naudet, and Hanlon), 192, 193
9/11, 1–3, 10–11, 12, 13, 15, 16, 18, 19, 20–22, 24, 25, 30, 57, 61–62, 84, 110, 169, 180–82, 183–85, 186–87, 188–90, 191–92, 193, 194, 196–97, 198–99, 200–201, 202, 203, 204, 206–209, 216, 217, 218, 219; falling bodies, 2, 3, 10–11, 12–13, 13n21, 181, 182, 186–87, 190–91, 192, 193, 194, 199, 201; representations of, 61, 192, 193, 194; texts after, 1–3, 5; 9–22, 24–26, 30, 57–58, 60–62, 84, 99, 110, 134–35, 166–67, 169, 180–94, 196–209, 216, 217, 218, 219; theory, 10–11, 19–22, 24–25; victims of, 2, 13. *See also* 9/11 Memorial; Ground Zero; terrorist attacks; vertical structures; verticality
9/11 Memorial, 1, 12, 14–20, 186; building of, 14–16, 18, 19, 20, 26, 57; composite, 18; Foundation Hall, 16; Last Column, 16, 18; Museum, 12, 14, 16, 18, 19; pools, 14, 15–16, 17, 18

acrobatics, 201. *See also* wire-walking
Adams, Henry. *See The Education of Henry Adams*
Afghanistan. *See* Gulf Wars
African Americans: in *Almanac of the Dead*, 118, 120, 121–23; in Winsor McCay's work, 50–52, 55. *See also* racism
afterwardsness. *See* nachträglichkeit
Alexie, Sherman, works of, 113–15, 141
Allix, Annie, 182, 184, 185, 186, 188, 198, 199, 205–6
Almanac of the Dead. See under Silko, Leslie Marmon, works of
American Airlines, 77. *See also under* terrorist attacks
American Indian Civil Rights Act, 121
Amexica, 128
amusement parks, 54–55, 57, 99–100
animation, 27, 68, 77–78
anticipation. *See* temporality
Anzaldua, Gloria, 130–31, 146
apocalypse: and atomic bomb, 148; California, 86–87; cityscapes, 56,

62, 201; in Ginsberg's work, 161–62, 167–69, 170–71; and oil production, 81; in Silko's work, 119–20, 122, 129–30, 136–37, 145; in Sinclair's work, 83–84, 87, 101–2, 130; post-9/11, 201. *See also* disaster

architecture, as metaphor, 13–14, 15, 19–20, 45–46, 58, 210–14. *See also* vertical structures; verticality

Armstrong, Neil, 154

atomic bomb. *See under* nuclear

Atomic Culture, 149

Auster, Paul, 181, 195–96, 199

aviation: and abstract verticality, 11; and concrete verticality, 2, 3–5, 29, 56, 99; in Ginsberg's work, 162–63; in Native American literature, 129, 137, 139–44; post-9/11, 2, 21, 181, 185, 193; and witnessing atomic detonation, 110, 165–66; and witnessing oil spill, 110. *See also* verticality; flight

awe. *See* sublime

Bakhtin, Mikhail, 44

Banita, Georgiana, 20–21, 22

Barthes, Roland, and tower form, 199, 205

Batista, Fulgencio, 157

Battersby, Christine, 26–27, 178, 206, 207, 208–9

Baudelaire, Charles, 33

Baudrillard, Jean, 11, 184

Beat movement, 150, 151, 152, 160

Benjamin, Walter, 76–77, 215

Bercovitch, Sacvan, 9

Bevis, William, and homing plot, 132–34, 136, 137

Beyond the Pleasure Principle (Freud), 42, 46, 48

Blondeau, Jean-Louis, 182, 190, 199

Blondin, Charles, 200–201

bombing. *See* nuclear: atomic bomb; *see also* terrorist attacks

borders: in Native American culture, 122–23, 141, 145–46; in Silko's work, 113, 118, 120, 122–23, 126–32, 134–36, 140, 145–46; U.S./Mexico border, 127, 128, 129, 130–31, 134–35, 138, 140, 146. *See also* cartography; frontier

Bosch, Hieronymus, 61

Boyer, Paul, 149, 166

Bredehoft, Thomas A., 45–46, 72

Brenkman, John, 42, 64

Brill, A. A., 41

Brown, Bill, 93, 99–100

Buddhism, 160, 169, 174–79. *See also* Heart Sutra; Diamond Truth

Building Stories (Ware). *See under* Ware, Chris, works of

Bukatman, Scott, 34, 36, 54, 68, 70–71

Burke, Edmund, 26–27, 206

Burtynsky, Edward, 93

Butler, Judith, 166–67

Calderon, Felipe, 128

California, 86–87, 91–92. *See also* apocalypse; frontier; horizontality

Carroll, John, 30

cartography: 2, 3, 28–29, 86–87, 91, 113–14, 115, 116–18, 121, 122, 125–29, 130–31, 132, 133, 134–36, 138–39, 141, 144, 145–46, 219; in *Almanac of the Dead*, 113, 114, 132, 116–18, 122, 125–29, 130–32, 133, 134, 136, 138–39, 141, 145–46; California, 86–87, 91; history of mapping North America, 2, 3, 28–29, 113–14, 115, 116–18, 121, 128, 129, 131–32, 133, 134–36, 145; Native American concepts, 113, 117, 127–28, 141, 144, 145–46

Caruth, Cathy, 23–24, 25, 191, 214

Castro, Fidel, 157

cathexis, 43

Ceremony. *See under* Silko, Leslie Marmon, works of

Chelsea, David, 184–85

Chester, George Randolph, 52

Chicago World's Fair, 3, 6, 15, 54

Chute, Hillary, 63, 72, 76

cinema, 1, 5, 9, 10, 24–25, 27, 30, 40, 55, 66, 70–72, 74, 75–78, 83–85, 93, 102, 106, 109, 133, 151, 180–209,

216, 218–19; city symphony films, 187, 202–4, 208, 209, 216; as distinct to other media, 66, 74, 75–78; emergence of, 9, 40, 70–72, 74, 75–77, 202–3; heist film genre, 187–88, 190, 193, 208; post-9/11, 1, 10, 18, 19, 180–209; theory, 24–25, 66, 70–72, 74, 75–78, 93, 183, 187, 193–94, 202, 203. See also *Man on Wire* (Marsh); *Smoke Signals* (Eyre); temporality; *There Will Be Blood* (Anderson)

city symphony films, 187, 202–4, 208, 209, 216

claustrophobia: and California, 56; in McCay's work, 38, 47, 49, in Native American literature and culture, 118–26, 133, 138–39; and New York, 28, 56, 57

Cold War, 149, 160

comics, 1, 27–28, 30, 33–40, 42, 43–44, 45–78, 109–10, 184–85, 210–14, 215, 216, 217, 218–19; as distinct to other media, 70–74, 75, 76–78; emergence of, 33–36, 54–55, 61, 70–72, 75, 77; theory, 45–46, 63–64, 66, 68, 70–72, 74, 75, 76–78. See also McCay, Winsor, works of; Ware, Chris, works of

communism, 155, 157

Coney Island, 54–55, 93–94, 99. See also amusement parks; disaster spectacles

Corso, Gregory, 151, 155

creative destruction, 183–84

Cubism, 35

Davis, Mike, 86

Dawes Act, 115–17, 120–22, 124–26

Dawes, Henry L., 115–17, 120–22, 124, 127, 135–36

Debord, Guy, 216, 218

de Man, Paul. See Man, Paul de

Deepwater Horizon. See under oil spills

DeKoven, Marianne, 72, 76n94

Deleuze, Gilles, 77–78

DeLillo, Don. See *Falling Man*

Derrida, Jacques, 21n39, 81, 182n4, 208

Diamond Truth, 175, 177. See also Heart Sutra; Buddhism

Dirks, Rudolph, 34

disaster, 56, 79–84, 86–87, 91–94, 99–100, 179, 214, 216, 218, 220; disaster culture, 28, 79–81; disaster spectacle, 54–55, 57, 93–94, 99–100; disaster theory, 79–82, 91, 214, 218; and narrative, 56, 80–94, 99–100, 122, 129–30, 146, 168–69, 179, 184, 201–2, 213, 214, 216, 218, 220; and oil, 28, 79–84, 86–87, 91–94, 99–100, 109; and post-9/11 texts, 10, 99, 201–2, 218; and the sublime, 27, 99–100, 110, 214, 218, 220. See also apocalypse; oil; vertical collapse

Doane, Mary Ann, 70–71, 77, 100, 215

Dream of the Rarebit Fiend (McCay), 27, 33–34, 36–38, 48–50, 52, 57–58, 69, 78

dream-comics, 32–33. See also McCay, works of

dreaming. See Freud, Sigmund: theories of dreaming; McCay, Winsor, works of

Dreamland (amusement park), 54–55

Drew, Richard, 13n21

drugs, as metaphor, 4, 48–49, 123–24, 139–41, 145

Education of Henry Adams, The (Adams), 5–10, 12, 32–33, 148, 161, 209

Eiffel Tower, 3, 15

Erdrich, Louise, 113, 132

exceptionalism, 18, 21–22, 182n4, 198–99, 204–5

Extremely Loud and Incredibly Close (Foer), 12–14, 19–20, 24–25, 191, 192, 205, 212

Exxon Valdez. See under oil spills

Eyre, Chris, 133

falling, 2–3, 4, 5, 10–11, 12–14, 13n21, 22–25, 26, 27, 33, 36, 38–40, 46–47, 53–54, 58, 60, 61, 62, 64, 66, 69, 74–75, 106–8, 109, 112, 142, 143, 161, 181, 182, 186–87, 190–92, 193, 194, 195, 199–200, 201, 205, 208, 214

falling bodies. See under 9/11

Falling Man (DeLillo), 191, 192

Fat Man. *See* nuclear: atomic bomb
fifth world, 113, 113n6, 120n19
fire. *See under* oil
fission. *See under* nuclear
flight, 3–5, 12, 13, 22, 27, 33, 36, 38, 46–47, 55, 80, 98, 99, 111–12, 114–15, 119, 123, 126, 129, 137–44, 146, 181, 186, 193; in Freud's work, 46–47; and *Man on Wire*, 181, 186; in McCay's work, 27, 33, 36, 38, 55; in Native American literature, 111–12, 114–15, 119, 123, 126, 129, 137–44, 146; post-9/11, 12, 13, 22, 193
Flight 93. *See under* terrorist attacks
Florey, Robert, 203
Foer, Jonathan Safran. See *Extremely Loud and Incredibly Close*
Ford, Henry, 88
Fortress Continent, 134–36
Foundation Hall. *See* 9/11 Memorial Museum
Freedom Tower. *See* One World Trade Center
Freud, Sigmund, 9, 25, 27–28, 40–48, 50n38, 60, 63: nachträglichkeit, 60; pleasure principle, 46, 48; theories of dreaming, 41–42, 44–47, 48, 63, 78; theories of ego and id, 43–44; theories of repression, 44–45, 47, 48, 51
Freud, Sigmund, works of: *Beyond the Pleasure Principle*, 42, 46, 48; *The Ego and the Id*, 44; *The Interpretation of Dreams*, 9, 41, 46–47
frontier, 2–3, 28, 29, 54, 55, 86–87, 88, 109, 114, 115, 122–23, 126–27, 128, 130–32, 136, 152–53, 186, 217–18; closure of, 2–3, 28, 54, 55–56, 88, 109, 114, 126–27, 217; as metaphor, 2–3, 54, 55–56, 86–87, 88, 109, 114, 217–18; as paradigm, 2–3, 29, 122–23, 128, 130–32, 136; Turner thesis, 3; vertical, 2, 54, 130–32, 136, 152–53, 186, 217–18

Gale Ferris Wheel, 3
General Allotment Act. *See* Dawes Act
Gerstein, Mordicai, 184n14
Gilbreth, Frank B., 70, 71

Ginsberg, Allen, 29–30; 108, 145, 147–79, 215, 216, 217, 219
Ginsberg, Allen, works of: "America," 151, 157; "Ballade of Poisons," 169, 171–72; "Howl," 150, 154, 172; journal fragments, 151–52, 155–56, 161–63, 164, 167–69, 174; "Nagasaki Days," 169–71; "Plutonian Ode," 169, 172–79; "POEM Rocket," 152–55, 156–59, 160, 164, 172, 178, 215; prose, 160–61, 164. *See also* nuclear: atomic bomb, in Ginsberg's work
golliwogg, 50–51. *See also* racism
gravity, 4, 13, 23, 24, 36, 140, 146
Great Chain of Being, 4
Great Year, 173
Greek mythology: Demeter, 50; Hades, 173; Hypnos, 50; Icarus, 5; Morpheus, 50; Persephone, 173; Phaeton, 5; Thanatos, 50
Greengrass, Paul. See *United 93*
Ground Zero, 14–20, 21, 26, 57, 180–83, 191–92, 198n34. *See also* 9/11 Memorial Museum
Groves, Leslie, 148
Gulf Wars, 84, 218. *See also under* oil spills

Habermas, Jürgen, 33–34
"Happy Hooligan" (Opper), 34
"Have We Failed with the Indian?" (Dawes), 115–16, 120–22, 124, 127, 135–36
Hearst, William Randolph, 34, 52
Heart Sutra, 175, 176–77. *See also* Buddhism; Diamond Truth
Heckel, Jean-François, 189, 199, 204, 205
heist film genre, 187–88, 190, 193, 208
Hine, Lewis W., 96–97
Hirsch, Marianne, 63
Hitchcock, Alfred, 204
"Hogan's Alley" (Outcault), 34
homing plot. *See* Bevis, William
horizontality, 2–3, 28–29, 54, 86–89, 90–91, 109, 114, 118–19, 122, 123–27, 128–29, 133, 136, 138–39, 141, 144, 146, 189, 202, 206, 217, 219

Houdini, Harry, 200–201
House Made of Dawn (Momaday), 112, 132
Hughes, Robert, 55, 57, 88
Husserl, Edmund, 179

illegality. See under *Man on Wire*
In the Shadow of No Towers (Spiegelman), 61–62, 169, 180
indexicality, 215
indian (critical term), 29, 114, 132n44
Indian Citizenship Act, 121
Indian Intercourse Acts, 117
Indian Removal Bill, 28–29, 114, 129
Indians. See Native American people
Iraq. See Gulf Wars
Ismoil, Eyad, 189. See also terrorist attacks: 1993 WTC attacks

Jackson, Andrew, 28–29
James, Henry, 184
James, William, 9, 40, 43, 45, 46
Jefferson, Thomas, 2
Jimmy Corrigan (Ware). See under Ware, Chris, works of
Julius, Anthony, 194
Jungle, The. See under Sinclair, Upton, works of
Jungle Imps, The (McCay), 52–53, 75–76
Just So Stories (Kipling), 52

Kant, Immanuel, 26, 166, 178, 199, 205, 206, 207, 216
Katzenjammer Kids, The, 34
Kaufman, Eleanor, 190–91
King Morpheus (*Little Nemo in Slumberland*), 38, 50, 58, 74
Kipling, Rudyard, 52
Klein, Naomi, 134–35
Knerr, Harold H., 34
Kohan, Jenji, 129

Langley, Samuel Pierpont, 6
Lasch, Christopher, 41–42
Last Column. See under 9/11 Memorial

Latinos/as, 118, 122, 135
Laurence, William, 165–67, 170, 178
LeMenager, Stephanie, 79–80, 81, 88n18, 94, 105
Lentricchia, Frank, 197, 198
Let the Great World Spin (McCann), 181n, 184n10, 197, 201–2
Lewis, Mark, 195
Little Nemo in Slumberland (McCay), 27–28, 33–36, 38–42, 43–55, 57–70, 72–78, 95, 99, 146, 215, 216, 217, 218–19
Los Angeles. See California
Luna Park, 54–55

Man on Wire (Marsh), 10, 30, 180–209, 216, 218–19; and city symphony films, 187, 202–4, 208, 209, 216; and heist film genre, 187–88, 190, 193, 208; and illegality, 186–90, 194–97
Man, Paul de, 23–24
Man Who Walked between the Towers, The (Gerstein), 184n14
Manhattan. See New York
Manhattan Project. See under nuclear
Manifest Destiny, 29
maps. See cartography
Marcus, Laura, 44, 74, 75
Marey, Etienne Jules, 70, 71
Marsh, James, 196–97, 208
Marx, Leo, 89, 97
masculinity, 86, 96–97, 102–3, 108, 178, 217
Mathews, John Joseph, 137. See also *Sundown*; *Talking to the Moon*
Mavroudis, John, 191
McAuliffe, Jody, 197, 198
McCann, Colum, 197, 201–2
McCay, Winsor, 27, 41, 53, 77, 78
McCay, Winsor, works of: *Dream of the Rarebit Fiend*, 27, 33–34, 36–38, 48–50, 52, 57–58, 69, 78; *The Jungle Imps*, 52–53, 75–76; *Little Nemo in Slumberland*, 27–28, 33–36, 38–42, 43–55, 57–70, 72–78, 95, 99, 146, 215, 216, 217, 218–19
McCloud, Scott, 63, 64, 66–68, 76n94

melodramatic didacticism, 80, 82
Melville, Herman. See *Moby-Dick*
memorialization. See 9/11; 9/11 Memorial
Meyerowitz, Joel, 181
Millennium: A Comedy of the Year, The. See under Sinclair, Upton, works of
Mitchell, William J. T., 51
Moby Dick (Melville), 103
modernism, 33–36, 40, 42–43, 44, 48, 54, 55, 57, 60, 62, 63–64, 69, 70–72, 75, 93, 108, 202–3, 208, 218
Momaday, N. Scott, 151. See also *House Made of Dawn*; *The Way to Rainy Mountain*
Mook, Delo E., 35
Moore, Jim, 195, 199, 205
Moore, Rowan, 16
Mumford, Lewis, 56–57
Muybridge, Eadward, 70–71

nachträglichkeit, 60
NAFTA. See Fortress Continent
Native American history: 28–29, 113–18, 120–22, 124–26, 126–27, 131–32, 134, 135–36, 138, 141, 145–46, 217; decimation, 28–29, 113–14, 126–27, 145–46, 217; displacement, 28–29, 112, 113–18, 120–22, 124–26, 126–27, 131–32, 134, 135–36, 138, 141, 145–46, 217; inequality, 28–29, 112, 113–18, 120–22, 126, 131, 131–32, 134, 135–36, 138, 145–46, 217
Native American peoples: Apaches, 127; Cheyennes, 113–14; Coeur d'Alene, 133; Creek, 123n27; Hopi, 113n6, 120n19; Kiowa, 124; Laguna Pueblo, 112, 123, 133–34, 136, 137; Sioux, 117
Native American theory: borders and frontiers, 113–114, 115, 118, 122–23, 126–27, 128, 129, 130–32, 134–36, 136, 138, 141, 145–46, 217; survivance, 29, 146; trickster, 29, 131
Nelson, Joshua B., 114–15
New York: and science fiction, 56, 184; as site of disaster, 56, 57, 183–84, 201–2; spatiality, 27, 28, 56, 57, 86, 183–84, 202. See also 9/11; 9/11 Memorial; verticality
Niagara Falls: as metaphor, 16, 100–101, 104; and Charles Blondin, 200
Nobel, Philip, 57
North Tower. See World Trade Center
nuclear: atomic bomb, 3, 21, 29–30, 80, 110, 129, 147–50, 151–52, 155, 158, 159–60, 161–62, 163, 164–6, 166–69, 170, 172, 178, 179, 215, 216, 218, 219; Fat Man, 165; fission, 149, 159–60, 161, 163, 173, 178, 179; in Ginsberg's work, 29–30, 148–49, 150–52, 153, 155–56, 158, 160, 161–63, 164–65, 166–69, 169–79, 215, 216, 218–19; Manhattan Project, 147; over Japan, 3, 110, 147, 148, 151, 161, 162, 165–66, 167, 168, 170, 178; technology, 29–30, 80, 147–49, 150, 151, 152, 153, 155–56, 158, 159–60, 161, 163, 167, 168–69, 170, 171, 172, 173, 174, 175, 178, 179, 215; Trinity test, 147–48, 150, 151
Nye, David E., 88, 95n29, 97, 100, 158, 199

oil: history of, 4, 28, 79–80, 84, 85, 86, 87, 94–96, 97, 98, 109–10; derricks, geysers, and rigs, 4, 28, 29, 79, 80, 81, 82, 83–84, 86, 87, 88, 92, 93, 94–97, 98, 100–101, 102–3, 104, 109, 114, 115, 144, 216, 219; documentaries, 81, 93; fire, 28, 79–82, 82–85, 86–87, 92, 93, 94, 99, 101, 104; production of, 28, 29, 79–82, 82–83, 84, 85, 86, 87, 91, 92, 93, 94–99, 100, 101, 102–10, 169, 179, 217
Oil!. See Sinclair, Upton, works of
oil spills, 81, 84, 110; Deepwater Horizon, 81, 82, 93; Exxon Valdez, 84; Gulf War spill, 84; Santa Barbara, 110
One World Trade Center, 1, 14–15. See also Ground Zero; 9/11 Memorial
op de Beeck, Nathalie, 214
Oppenheimer, J. Robert, 147
Opper, Frederick Burr, 34

INDEX | 253

Orlovsky, Peter, 155, 169
Ortiz, Alfonso, 122–23
Ortiz, Simon J., 113
Outcault, Richard F., 34
outer space, 29–30, 80, 130, 152–59, 170, 178, 215; Space Race, 29–30, 153, 154, 155, 158, 172, 178; space travel, 80, 153, 154, 155, 156, 157, 158, 159, 170, 172, 178, 215
Owens, Louis, 127, 131–32, 144

Page, Max, 56
Paris Exposition, 3, 6, 7
Pearl Harbor, 166, 170
Pendakis, Andrew, 94
Petit, Philippe, 10, 11, 30, 110, 180–209, 217
photography, 1, 9, 13, 25, 55, 70–71, 72, 77, 93, 95, 96–97, 180–81, 185, 189, 190, 193, 195, 202, 218; Abu Ghraib, 218; emergence of, 70–71
Plainview, Daniel (*There Will Be Blood*), 83–85, 106
Plank, Max, 9
pleasure. See sublime
post-9/11 narrative. See 9/11, texts after
Pozorski, Aimee, 10–11, 12, 13n21, 20–21, 22n43, 23–24
protest: against nuclear technology, 160, 170, 174, 178; against war, 160; narrative as, 30, 152, 170, 174, 178, 218–19
psychoanalysis, (development of), 9, 27, 34, 40–48, 50, 60, 76–77, 78, 103. See also Freud, Sigmund
Pulitzer, Joseph, 34

Quam-Wickham, Nancy, 87

Rabi, Isidor, 148
racism: early twentieth century, 50–54, 55, 58, 60; governmental/systematic, 118, 121–23; representations in *Almanac of the Dead*, 129–30
radioactivity. See nuclear: atomic bomb
Relis, Paul, 110

retroactivity, 62, 84, 85, 191–92, 201–3, 207–9, 216. See also nachträglichkeit; temporality
Rifkin, Mark, 121
Rinpoche, Chogyam Trungpa, 169
Roeder, Katherine, 34, 55, 214
Ross, Bunny (*Oil!*), 83, 85–86, 88, 89, 91–92, 96, 97–108, 110
Ross, J. Arnold (*Oil!*), 84, 85, 88–89, 96–97, 102, 105–7, 108
Ross, Luana, 122
Rothberg, Michael, 24, 25
Ryan, Judith, 42

Sabin, Paul, 87
San Francisco Renaissance, 150, 164
Scheeler, Charles, 202
"The School Days of an Indian Girl" (Zitkala-Sa), 124–26
Senk, Sarah, 15, 19
sexual imagery, 43, 102–4, 108–9
"Significance of the Frontier in American History." See frontier: Turner thesis
Silko, Leslie Marmon, 113, 123, 131, 151
Silko, Leslie Marmon, works of: *Almanac of the Dead*, 28, 111–13, 115–20, 122–46, 152, 155, 217; *Ceremony*, 111–12, 113, 118, 124, 127, 132, 133, 138, 139, 141
Simpson, David, 16, 21, 22
Sinclair, Upton, 82, 89, 97, 98, 103–4
Sinclair, Upton, works of: *The Autobiography of Upton Sinclair*, 98, 103; *Candid Reminiscences: My First Thirty Years*, 89; *The Jungle*, 82, 101–2; *The Millennium: A Comedy of the Year* 2000, 130; *Oil!*, 28, 80–110, 114, 144, 148–49, 169, 179, 217, 219
Six Gallery, 150
Skyscraper Symphony, 203–4
skyscrapers. See under vertical structures; see also verticality
Slotkin, Richard, 152
Slumberland, 42, 46, 48, 51–52, 53, 54–55, 68–69, 69–70, 74
Smith, Owen, 191

Smoke Signals (Eyre), 133
Snyder, Gary, 108–9, 150
"Soaring Spirit" (Mavroudis and Smith), 191–92
South Pool. *See* 9/11 Memorial Museum
South Tower. *See* World Trade Center
Space Race. *See* outer space
Spanish-American War, 52, 116n16
spectacle. *See under* disaster
Spiegelman, Art, works of: *In the Shadow of No Towers*, 61–62, 169, 181; *Maus*, 61
Sputnik, 153, 154
Strand, Paul, 202
Stauffer, John, 152
Stockhausen, Karlheinz, 198
Sturken, Marita, 14, 16, 18–19, 20, 183
sublime, 20, 26–27, 30, 62, 82, 93, 94–95, 97–98, 100–101, 110, 144–45, 148–49, 150, 152, 158–60, 162, 164–65, 165–66, 178–79, 199–200, 201, 202, 205–9, 214–15, 216, 217, 218, 219, 220. *See also* Burke, Edmund; Battersby, Christine; Kant, Immanuel; technological sublime
Sullivan, Louis H., 56–57, 216–17
Surrounded, The (McNickle), 113, 132–33, 137
survivance, 29, 146

Talking to the Moon (Mathews), 113, 114, 137–38, 144
tall buildings. *See* vertical structures
Taylorism, 70
technological sublime, 100–101, 148, 199
temporality, 6–10, 11, 12–13, 14, 19, 21–22, 24–26, 27, 30, 31, 32, 33, 35, 47, 57–58, 60, 62–78, 99–100, 107, 110, 111–12, 113n6, 114–15, 122, 127, 138, 139–40, 141, 143, 145, 146, 148, 156, 160–62, 166, 173, 175, 176, 179, 182–83, 185, 186, 191, 193, 201–2, 206–9, 211, 213–16, 217, 218–20
terrorism, theory of, 194–95, 197–99
terrorist attacks: 1993 WTC attacks, 189–90; American Airlines 77, 21; United Flight 93, 21, 193

There Will Be Blood (Anderson), 83–85, 102, 106, 109
Tillett, Rebecca, 130–31
time. *See* temporality
To Reach the Clouds (Petit), 10, 181, 204, 206
Trachtenberg, Alan, 90, 97
trauma, 3, 4, 11, 13–14, 20–26, 60, 112, 143–44, 182–83, 190–91, 206, 207, 208, 214, 218; theory, 11, 14, 20–25, 60, 183n8, 190–91, 206, 207, 208, 214, 218; traumatic repetition, 20
tunnels, 4, 128
Turner thesis. *See under* frontier
Twin Towers. *See under* vertical structures

United 93 (Greengrass), 192, 193, 194. *See also* terrorist attacks: United Flight 93
Upton, Florence Kate, 51

Vargish, Thomas, 35
vertical axis, 4–5
vertical frontier. *See under* frontier
vertical structures: derricks, geysers, and rigs, 4, 28, 29, 79, 80, 81, 82, 83–84, 86, 87, 88, 92, 93, 94–97, 98, 100–101, 102–3, 104, 109, 114, 115, 144, 216, 219; skyscrapers, 2, 3–4, 5, 13–16, 18, 19, 27, 28, 29, 30, 33, 36, 40, 46, 54, 56, 55–57, 58, 59, 60, 62, 80, 82, 95–96, 96–97, 114, 115, 130, 144, 161, 162, 182, 183–84, 186, 191, 199, 201–2, 203, 204, 205, 209, 210–14, 216–17, 219, 220; skyscrapers, history of, 3–4, 5, 15, 28, 40, 54, 55–57, 95, 183–84; World Trade Center (Twin Towers), 2, 3, 10, 11, 12, 13, 14, 15, 18, 19, 26, 30, 61, 110, 169, 180–84, 185, 186, 187, 188–89, 190–92, 193, 195, 196, 197, 198, 203–4, 205–6, 208, 217. *See also* 9/11 Memorial; verticality, vertical collapse
verticality: and anxiety, 1, 8, 11, 12, 19, 20, 22, 26, 30, 32, 35–36, 46–47, 50–51, 53, 57, 58, 60, 61, 62, 68, 69–70, 74, 78, 82, 83, 85, 86, 91, 94, 97, 100, 109, 146, 153, 168–69, 188, 189–92, 194, 201, 208, 209, 213,

214, 218; falling, 2–3, 4, 5, 10–11, 12–14, 13n21, 22–25, 26, 27, 33, 36, 38–40, 46–47, 53–54, 58, 60, 61, 62, 64, 66, 69, 74–75, 106–8, 109, 112, 142, 143, 161, 181, 182, 186–87, 190–92, 193, 194, 195, 199–200, 201, 205, 208, 214; flight, 3–5, 12, 13, 22, 27, 33, 36, 38, 46–47, 55, 80, 98, 99, 111–12, 114–15, 119, 123, 126, 129, 137–44, 146, 181, 186, 193; vertical collapse, 2–3, 10, 11, 12, 14, 19, 24, 25, 26, 27, 30, 33, 36, 40, 46, 48, 51–52, 54, 56, 57–58, 58–60, 61, 62, 82, 169, 180, 181, 182–83, 183–84, 185, 187, 188, 189, 191, 192, 196, 197, 198, 199, 201–2, 203, 204, 208, 216

verticality, theory: abstract and concrete, 3–4, 5, 11, 12, 216–17, 219–20; emergence, 3, 4–10, 12, 15, 22; etymology, 8, 13; existing studies, 22

Vietnam War, 30, 149, 160, 170, 171

Vizenor, Gerald, 29, 114, 132n44, 145, 146

Walonen, Michael K., 80

Ware, Chris, 63, 68, 72

Ware, Chris, works of: *Building Stories*, 30–31, 210–14, 217; *Jimmy Corrigan*, 212

Washington Monument, 3

Way to Rainy Mountain, The (Momaday), 113, 124

Welch, James, 113

Whissel, Kristen, 5

"The White Man's Burden" (Kipling), 52

Whitman, Walt, 150, 162, 163, 164, 174

Williams, William Carlos, 108, 150

Wilson, Rob, 159, 214–15

wire-walking, as metaphor, 181, 195–96, 199. See also *Man on Wire*

Womack, Craig: *Drowning in Fire*, 141, 143; Native American theory, 123n27

Woolf, Virginia, 27, 40, 43

World Trade Center. See *under* vertical structures

Wounded Knee Massacre, 127

Wright brothers, 3

yellow journalism, 34, 52

Yellow Kid (Outcault), 34

Yousef, Ramzi, 189. See also terrorist attacks: 1993 WTC attacks

Zapatista Army of National Liberation of Chiapas, 128

Zeman, Scott C., 161

Zitkala-Sa, 124–26

Žižek, Slavoj, 184

www.ingramcontent.com/pod-product-compliance
Lightning Source LLC
Chambersburg PA
CBHW021139230426
43667CB00005B/181